KEPLER'S DREAM

KEPLER'S DREAM

BY JOHN LEAR

WITH THE FULL TEXT AND NOTES OF

Somnium, Sive Astronomia Lunaris,
Joannis Kepleri

TRANSLATED BY PATRICIA FRUEH KIRKWOOD

UNIVERSITY OF CALIFORNIA PRESS
BERKELEY AND LOS ANGELES 1965

University of California Press
Berkeley and Los Angeles, California
Cambridge University Press
London, England
© 1965 by The Regents of the University of California
Library of Congress Catalog Card Number: 64-21775

Dedicated to Dorothy

ACKNOWLEDGMENTS

Grateful acknowledgment is made to Mrs. Patricia Kirkwood for her translation of Johannes Kepler's *Dream* from Latin into English; to Mr. Norman Cousins, editor, and to Mr. J. R. Cominsky, publisher, of *Saturday Review* for permission to reproduce in this book passages from that translation which were originally published in *Saturday Review*; to Dr. C. Doris Hellman for reading and criticizing the early drafts of the book manuscript; to Mrs. Mary Virginia Kahl and to Mr. R. Y. Zachary, of the University of California Press, for editorial advice and assistance; to Dr. M. W. Makemson, University of California astronomer, for suggestions and sketches; and to my wife, Dorothy, for meticulous proofreading.

J. L.

CONTENTS

INTRODUCTION

and

INTERPRETATION

I

Voyaging from earth to other planets of the sun became a reasonable subject for intellectual speculation in the year 1609. That was when Johannes Kepler, mathematician to the Hapsburg Emperor Rudolph II in Prague, published two of his three great discoveries about the orbits of the planets. Most people of that time still believed that the sun circled the earth once every twenty-four hours; only a few thought that the earth revolved on its own axis while traveling around the sun, and they were of the opinion that the earth followed a circular path. Kepler plotted that path through the sky in relation to the more irregular path of the planet Mars (the positions of which had been recorded periodically over several decades with unprecedented accuracy by the Danish astronomer, Tycho Brahe) and saw that neither path was a circle; both were ellipses. He also saw that the speed with which a planet traversed its ellipse varied in a regular pattern, accelerating with approach to the sun and decelerating with departure from the sun. By refining the mathematics he used to determine those movements, Kepler could tell where any planet would be in relation to any other planet at any chosen moment, and consequently in what direction and how far an expedition from one planet to another would have to proceed. Prior to his discovery of the navigational realities, any interplanetary expedition could have arrived at its destination only by the wildest luck.

Three hundred and fifty-three years would have to pass before metals could be alloyed for spaceship hulls, fuels developed for propulsion of those hulls, and electric controls devised for guidance of an actual interplanetary passage from earth to Venus by the robot named Mariner II. But Kepler began to elucidate the scientific problems entailed in such voyages in the same year in which his laws of planetary motion were printed. In the summer of 1609 he wrote out a plan for landing on the moon, earth's closest neighbor in the sky.

Kepler mentioned his project in the scientific literature for the first time in an open letter to his Italian contemporary, Galileo Galilei, on April 19, 1610. The letter was a reply to the historic pamphlet, *Message from the Stars*,[1] in which Galileo described what he had seen in searching the heavens through a telescope: mountains on the moon, thousands of stars theretofore unseen by anyone, and alongside the planet Jupiter four small luminous globes that changed places from one night to the next.

Galileo had published his *Message* in the spring of 1610, and had sent a copy across the Alps to Kepler. Although the science advisor to Emperor Rudolph did not yet possess a telescope through which to confirm Galileo's observation, there soon went out from the court printer in Prague a *Conversation with the Star Messenger*.[2]

In this *Conversation* Kepler reminded Galileo that the presence of mountains on the moon had been deduced long ago. The evidence for their existence had been recorded by Plutarch, the Greek essayist, fifteen hundred years before. Kepler himself,

. . . . . .

[1] Galileo Galilei, *Siderius Nuncius* (Venice, 1610), translated by Stillman Drake in *Discoveries and Opinions of Galileo* (New York: Doubleday-Anchor, 1957).

[2] Johannes Kepler, *Dissertatio cum Nuncio Siderio* (Prague, 1610); excerpts translated from *Joannis Kepleri Astronomi Opera Omnia*, edited by Christian Frisch (Frankfort, 1870), II, 288–311.

as a student of astronomy professor Michael Maestlin at
Tübingen University, had learned to estimate elevations on
the lunar surface by measuring shadows on the moon—the
technique the ancient Greeks had used. In 1593 the student
Kepler had written out a series of speculations derived from
these observations.

"Last summer," he confided to Galileo, the manuscript begun in 1593 had been expanded into "a complete geography of
the moon." [3]

Kepler's letter continued with a question:

"Who would have believed that a huge ocean could be
crossed more peacefully and safely than the narrow expanse of
the Adriatic, the Baltic Sea or the English Channel?"

Yet Christopher Columbus had proved the Atlantic safe for
voyaging little more than a century earlier.

Kepler went on to make a prophecy:

"Provide ship or sails adapted to the heavenly breezes, and
there will be some who will not fear even that void [of interplanetary space]. . . . So, for those who will come shortly to
attempt this journey, let us establish the astronomy: Galileo,
you of Jupiter, I of the moon." [4]

II

In the course of this rambling communication, Kepler passed
along to Galileo what seemed to be a piece of innocuous gossip. The imperial mathematician said he had not composed the
lunar geography on his own initiative; he had written the work
"to please Wackher."

Wackher was John Matthew Wackher von Wackenfels, a
distant relative of Kepler.[5] But Kepler did not mention the family relationship to Galileo, emphasizing instead that Wackher
was an ecclesiastical adviser to Emperor Rudolph II.

. . . . . .

[3] *Opera Omnia*, II, 297–298.
[4] *Ibid.*, p. 305.
[5] Max Caspar, *Kepler*, translated and edited by C. Doris Hellman
(London, New York: Abelard-Schuman, 1959), p. 161.

Now why should Wackher's official title have been drawn into the first disclosure of the lunar geography's existence? What need had Kepler, himself a favorite of the Emperor, for the sponsorship of an imperial court referee in church affairs? What weight would Wackher's name carry in matters of science?

Since Kepler himself never offered or suggested an explanation, it is reasonable to suppose that he thought none was required. He considered the facts too well established to need further repetition or elaboration. And during his lifetime they were. Everyone who mattered to him knew that he had studied for the Lutheran clergy but had never been assigned a pulpit; instead he had been shunted into the teaching of mathematics at a Lutheran seminary in Graz immediately after his graduation from Tübingen University. Ever after that, he had wrangled with the Lutheran authorities.

In the centuries following his death the story that came to be generally accepted was that the Lutheran fathers did not trust Kepler enough to accept him as a preacher. He was much too lenient with Calvinists and Catholics in questions of dogma. That was why he had to go into mathematics, and it was the cause of the bitterness between him and formal Lutheranism throughout his life.

I repeat: That is the commonly accepted story. But the late Max Caspar, the German historian who wrote the definitive biography of Kepler, rejected it. Kepler certainly did hold views that were considered heretical by the Lutheran hierarchy. He was resented because of them as long as he lived. However, Caspar insisted, those views were "not disclosed . . . to the servants of the church" prior to Kepler's acceptance of the mathematics teaching post at Graz. Hence they could not possibly have influenced the Graz assignment, the main turning point of Kepler's career. "Kepler was sent to Graz," Caspar concluded, "because, on the basis of his mathematical and astronomical knowledge, he was by far the most suitable candidate for the teaching position there, the only one worthy of consideration and likely to bring honor to Tübingen University."

I find an interesting shift in emphasis within a pair of remarks Caspar made in leading up to his statement of this conclusion. He wrote: "The Tübingen professors may well have shaken their heads, when they heard the young zealot so enthusiastically supporting Copernicus. Yes, they may also have gotten wind of his doubts [on certain points in Lutheran dogma]." [6] In other words, the Lutheran authorities may have heard of Kepler's divergent religious opinions from his intimates, but they undoubtedly had heard of his Copernican views, the sole question on this latter point being the extent of the formal expression of Lutheran disapproval.

To me, there is a significance in this distinction which has not been remarked on by others. Not only did Caspar draw a subtle line between Kepler's theological troubles and his scientific troubles, but the scientific dilemma was stated first as a matter of fact in contrast to the mere possibility of dogmatic wrangling.

How had Kepler's early enthusiasm for the Copernican theory of sun-centered astronomy been made so offensively obvious?

As Galileo was told in the *Conversation* of 1610, it had happened in 1593. While still a student at Tübingen, Kepler had written a dissertation about the moon. The purpose had been to demonstrate the simultaneous motion of the earth on its axis and around the sun. One of Kepler's classmates, Christoph Besold, a law student, had drawn from this lunar paper a series of theses that he proposed to develop in a public debate under University auspices. [7]

The proposal had been presented to the Lutheran authorities at Tübingen for approval. They had vetoed the debate. [8] Only one member of the faculty, astronomy professor Maestlin—Kepler's favorite teacher—supported Nicolaus Copernicus' reversal of the two-thousand-year-old notion that the sun revolved

· · · · · ·

[6] Caspar, *op. cit.*, p. 52.

[7] *Ibid.*, p. 48.

[8] Kepler's second footnote to the lunar geography says the debate would have been held *if* the appropriate official at Tübingen had approved.

around the earth.[9] All the rest of the Tübingen staff took the position that Martin Luther, founder of Lutheranism, and his scientific advisor, Phillip Melanchthon, had taken: that to accept Copernicus was to reject Holy Writ.[10] Had it not been the sun that the Jewish prophet Joshua commanded to stand still during the famous battle for Gibeon?

A few years after that suppression of his attempt to argue a minority opinion, Kepler was subjected to another act of Lutheran censorship. This time the blow fell on his first book, *Cosmic Mystery.*

In this work of youthful exuberance, Kepler had repeated Maestlin's teaching that the earth and the moon were made of similar stuff. He had also hinted at the existence of an attraction between them: "The moon follows, or rather is drawn, wherever and however the earth moves along." And he had tried to document the Copernican theory by establishing a fixed relationship between the distance of a planet from the sun and the time required for that planet to complete a circuit of the sun. The method he employed was to bound the apparent motion of the six planets up to that time discovered—Mercury, Venus, Mars, Jupiter, Saturn, and the earth—by separating them with the five regular solids of Euclidian geometry and then placing this one-within-another arrangement of planets and solids nestlike, with the sun in the middle of the nest.[11]

. . . . . .

[9] The earth-centered tradition bore the name of the Greek geographer Claudius Ptolemy. However, its actual father was a predecessor of Ptolemy, the astronomer Hipparchus, who knocked down a theory proposed by Aristarchos of Samos in the third century B.C. Aristarchos argued that the earth revolved on its axis while traveling around the sun. Just as Aristarchos' ideas could not stand against the challenges of Hipparchus, so the Copernican revival of Aristarchian thought would not have been able to stand without Kepler's introduction of the concept of elliptical orbits. See George Sarton, *A History of Science: Hellenistic Science and Culture in the Last Three Centuries B.C.* (Cambridge, Mass.: Harvard University Press, 1959), pp. 56–60.

[10] Caspar, *op. cit.,* p. 46.

[11] *Ibid.,* pp. 60–71.

Except for Professor Maestlin, who objected to the *a priori* nature of the orbits, the Tübingen faculty (which oversaw affairs at the Graz seminary) raised no objection to *Cosmic Mystery* on astronomical grounds. But before they would approve publication of the book they required Kepler to remove his original opening chapter, which was a detailed refutation of all the then-circulating arguments that depended on Biblical quotations to discredit the Copernican theory.[12]

After *Cosmic Mystery* appeared in 1597, Kepler became dissatisfied with the inexact match between the geometrical boundaries he had fixed *a priori* for the planetary orbits—an octahedron between Mercury and Venus, an icosahedron between Venus and the earth, a dodecahedron between the earth and Mars, a tetrahedron between Mars and Jupiter, and a cube between Jupiter and Saturn—and the actual positions the planets were seen to occupy in the sky from night to night. To bring his theory closer to reality, he decided to check the dimensions of his fixed geometrical orbits with the finest astronomical observations he could find.[13] There was no doubt of the identity of the observer he would turn to. Tycho Brahe, the Dane, was famous everywhere for his unexcelled precision. So Kepler sought out Tycho and joined him as junior partner in astronomical research in Prague in the year 1600.[14]

Following Tycho's death in 1601, Kepler worked out legal arrangements with the heirs which permitted mathematical analysis of Brahe's observational data.[15] The calculations continued for four more years before the imperial mathematician abandoned his earlier belief that the orbits of the planets could be described by circles of sizes determined by intervening shapes of geometrical solids. Finally forced to choose between

. . . . . .

[12] *Ibid.*, p. 68.

[13] *Ibid.*, p. 87.

[14] The partnership was sealed by a two-year contract.

[15] It was a very troubled arrangement, partly because Brahe, who could not accept the Copernican doctrine because he could not confirm it by observing the parallax of the earth, pleaded with Kepler not to use his data to support Copernicanism; but in the end it worked out for the advancement of knowledge. Caspar, *op. cit.*, pp. 121–123.

what tradition said ought to be true and what Brahe's careful observations said was in fact true, Kepler declared his faith in the experimental method by announcing that the orbits were not circles but ellipses.[16]

This one discovery revolutionized astronomy. Before it, everyone assumed that the planets had to travel in circles because the circle was accepted as the perfect geometrical form and because perfection was accepted as the normal state of heavenly affairs.[17] The necessity for accurate prediction of the alternating seasons of the year had required some reasonably satisfactory description of the pattern of movement of the heavenly bodies, but no one had ever been able to work out a system of circles that would perform as an actual working model of the planets and the stars. Kepler's ellipses introduced the missing dynamic element. Given elliptical orbits, the sun-centered system of planetary motion could be seen to work with the elegant simplicity of a well-built clock.

So little about this elliptical system was Copernican[18] that Kepler would have been justified in christening it with his own name. But whether as a stratagem to escape the full weight of theological opposition or as an expression of genuine modesty, the imperial mathematician chose to designate his radical innovation as a mere refinement of Copernicus' thought.[19]

Along with the advantages inherent in this self-effacing ap-

. . . . . .

[16] Although publication did not occur until 1609, Kepler disclosed the discovery in a letter in 1605.

[17] See various citations by Max Jammer in *Concepts of Mass* (Cambridge, Mass.: Harvard University Press, 1961), p. 53.

[18] Sarton, *op. cit.*, p. 297.

[19] "I do not understand at all which book you call the Copernican, for all my books are Copernican," Kepler wrote in a letter to Emperor Matthias' court physician, Johannes Quietanus Remus, on August 4, 1619. *Johannes Kepler—Gesammelte Werke*, edited by Walther von Dyck and Max Caspar (Munich: C. H. Beck'sche, 1955), vol. 17, letter 846, p. 364. Quoted in Carola Baumgardt, *Johannes Kepler: Life and Letters*, with an introduction by Albert Einstein (New York: Philosophical Library, 1951), p. 141. Also see Caspar, *op. cit.*, p. 298.

proach, Kepler acquired a problem. It dated back to Copernicus. As a functionary of the Roman Catholic Church in six-
teenth-century Poland, Copernicus understood the difficulty
the Christian clergy would face in interpreting a change from
an earth-centered to a sun-centered universe. For hundreds of
years the faithful had been taught to accept the Biblical tradition that God had created the earth as a home for man, and
that man's happiness in that home was God's foremost concern.
If now the earth should suddenly be reduced to a wandering
dependent of the sun, all the rest of Christian doctrine might
be threatened with doubt among the ignorant.

For a long time Copernicus could not resolve this conflict to
the satisfaction of his conscience. Accordingly, he refused to
publish his theory of sun-centered astronomy until after a German Lutheran, George Joachim von Lauchen, assimilated the
idea during a two-year visit with the Pole and published the gist
of it in 1540 in a letter signed with the pseudonym Rheticus.[20]
Even when, encouraged by the interest aroused by the Rheticus
letter, Copernicus finally did agree to release the whole of his
detailed argument, in *Revolutions of the Heavenly Spheres*, the
book was dedicated to Pope Paul III to discourage theological
criticism.[21]

The dedication had the desired effect. Catholic opinion at
first was relatively undisturbed.[22] The opening attack on Copernicus, who died just after his book went to press in 1543,
came from Luther and Melanchthon.[23] In an effort to disarm
this overpowering opposition from his own church, Andreas
Osiander, a Lutheran clergyman who supervised the printing
of Copernicus' book, removed its author's original introduction
and substituted an anonymous foreword. Whereas the introduction had presented the Copernican theory forthrightly, the
new foreword said the book meant only to describe the appear-

. . . . . .

[20] Edward Rosen, *Three Copernican Treatises* (New York: Dover,
1959), 2d edition revised, p. 10.
[21] Rosen, *op. cit.*, p. 27.
[22] Caspar, *op. cit.*, p. 22.
[23] *Loc. cit.*

ances of the heavens and should not be read to mean that the earth actually moved around the sun.

In publishing his discovery that the real orbits of the planets were ellipses, Kepler could not logically claim to be demonstrating the workability of the Copernican theory as long as the foreword to Copernicus' own book implied that Copernicus himself did not intend the theory to be taken literally. Therefore, the spurious foreword to *Revolutions of the Heavenly Spheres* had to be exposed. To Kepler's contemporaries, one of the biggest surprises in his second book, *New Astronomy: A Description of Celestial Physics, with Comments on the Motions of Mars,* published in 1609, was the revelation that the Lutheran Osiander had totally misrepresented the Catholic Copernicus' belief in the heliocentric motion of the planets.[24]

The Lutheran Church could hardly have been expected to rejoice over Kepler's insistence on broadcasting the true beliefs of Copernicus, especially not after the Tübingen authorities twice before had shown him the correct attitude to take on Copernicanism. What reaction might have been expected among the Lutherans, then, were it to become known that Kepler planned to top the exposure of Osiander by circulating, within the span of a few months, still another manuscript, a lunar geography, based on the very debating theses that had been bannned by the Lutheran authorities at Tübingen in 1593?

Kepler was, in fact, reluctant to write the lunar geography on that account. He said as much in an early passage of the geography—to those who could decipher the double meanings in which the lunar manuscript abounds.[25] He explained, in the same vein of double-talk in that same passage, that he finally wrote the work only after becoming persuaded that opposition to Copernicanism had become "sufficiently extinct." [26]

Too late he discovered that his earlier doubts had been justified after all. During the year 1609 he had hinted to the Duke

. . . . . .

[24] *Opera Omnia,* III, 136.
[25] Kepler's footnote 3.
[26] *Loc. cit.*

of Württemberg that a post on the Tübingen faculty would be welcome, and in 1611 he made a specific application for such a job. The application was rejected. According to Caspar, there were two reasons for this Tübingen slap. Only one of these is usually recalled: "He would certainly inspire the youth with the poison of Calvinism." The second reason was Kepler's advocacy of the Copernican theory. To put it, again, in Caspar's words: "And since he is 'an opinionist in philosophy [science],' he would otherwise [than in questions of dogma] also arouse much unrest at the University." [27]

Though not easily intimidated, Kepler did have a deep fear of censorship, which was to come to the surface when the first part of his *Epitome of Copernican Astronomy* was placed on the Roman Catholic Index (of books banned to Catholic readers) in 1619. His reaction then was to be to imagine pending demands that he renounce astronomy as a career in spite of his eminent position.[28] It may be supposed that some of this attitude affected him at the time he wrote the lunar geography, for he had just experienced trouble in connection with distribution of *New Astronomy*. The reasons for these difficulties do not survive in the historical record; however, Emperor Rudolph ordered Kepler to "give no one a copy" of that book "without our previous knowledge and consent." [29]

Therefore, everything considered, it made excellent sense for

· · · · · ·

[27] Caspar, *op. cit.*, p. 205.
[28] *Ibid.*, p. 298.
[29] Caspar, *op. cit.*, p. 141, had to guess about the motive. "Evidently," he said, the Emperor felt justified in restricting the sale because Kepler did the work on Hapsburg time; besides, Caspar added, there seemed to be a particular importance in the book that caused Rudolph to prefer a circulation list of his own choosing. My hypothesis is that Rudolph's sense of import was related to the politico-religious crisis that ultimately lost him his crown. Kepler describes this crisis briefly in the opening paragraph of the lunar geography. Caspar points out that Rudolph's control over the situation deteriorated to such a point that Kepler finally took matters over and sold the whole edition of the *New Astronomy* to the printer.

Kepler to advertise to Galileo and other scholars the involvement of Wackher von Wackhenfels, ecclesiastical referee at Emperor Rudolph's court, in the lunar geography's preparation. Any document prepared at the instigation of a religious arbiter in the imperial entourage would be examined very cautiously and pondered well by other theologians before they attacked it. Perhaps more importantly, Wackher's name and title would give second thoughts to court politicians who might otherwise be tempted into intrigue by the anti-Copernican clergy.

On his part, Wackher, being a convert from Lutheranism to Catholicism,[30] knew the treacherous politics of the religious situation in seventeenth-century Germany from personal experience. Certainly he had not reached his position of influence through excursions in rashness. Common prudence must have required him to attach provisos to publication of his name as the initiator of the lunar geography. Hence it is not surprising that the moon guide was designed for distribution to a restricted audience. Phrased in Latin, the international language of the learned, it was cast in the form of an allegory, the hidden meanings of which would be familiar only to scientists. As Kepler explained later in a letter to his friend, Matthias Bernegger, the text of the geography had been deliberately strewn with "almost as many problems as there are lines;" Kepler had expected these problems to be identified and worked out by readers competent to do so—"some by astronomy, some by physics, some by history." [31]

. . . . . .

[30] Caspar, *op. cit.*, p. 161.

[31] Written in 1623, this letter is quoted by Christian Frisch in his preface to the Latin text of the *Dream* in *Opera Omnia*, Vol. VIII: "I began to revise my *Lunar Astronomy*, or rather to elucidate it with notes, two years ago when I first returned to Linz. In my essay, there are as many problems as there are lines, which must be solved partly by means of astronomy, partly by physics, partly by history. But what would you? How many people would consider it worth the trouble to solve them? Men want amusements of this sort to be presented with a gentle arm, as they say, and are not prepared to wrinkle their brows over a game. Therefore, I decided to solve all the problems by means of successive notes following the text."

So far as I know, the events just described have not been presented before in the pattern in which they have been arranged here. Nor has anyone else assigned these events the values which seem to me implicit in them. It can be argued that in naming Wackher von Wackhenfels as the inspiration for the lunar geography, Kepler was merely acknowledging once again an encouraging influence he had acknowledged in previous works. And the allegorical form may be dismissed as simply another expression of the imaginative style Kepler had also employed at another time in writing, for Wackher's amusement, a scientific explanation of the shape of a snowflake. For me, however, there is an enormous gulf between a delightfully gay description of snowflakes, in which everything sparkles clearly on the surface, and the labyrinthine construction of double meanings which underlies Kepler's geography of the moon. I cannot escape the conclusion that this allegory, involving so much intellectual exercise by the imperial mathematician at a most trying period of his life, was written at a much deeper level than that of an offhand game. To clinch my case, I now call attention to the peculiar historical setting in which Kepler himself placed the lunar geography.

III

Why, it will be asked, should Kepler have worried about suppression of his work by the Lutherans while he lived in the shelter of the official family of the Catholic Emperor Rudolph II? If my reconstruction of the circumstances is right, the reason was that Rudolph's brother, the ambitious Archduke Matthias, had allied himself with Protestant forces in 1608 to strip Rudolph of authority over all the Hapsburg lands except Bohemia and neighboring Silesia.[32] Kepler went out of his way to state the equivocal situation at court in the opening lines of the lunar geography:

"In the year 1608, when quarrels were raging between the brothers, Emperor Rudolph and Archduke Matthias, people were comparing precedents from Bohemian history. Caught up

. . . . . .
[32] Caspar, op. cit., p. 187.

by the general curiosity, I applied my mind to Bohemian legends and chanced upon the story of the heroine Libussa, famous for her magic art."

Now Libussa was a sorceress only in the political sense. Folklore said she founded the ancient kingdom of Bohemia. By singling her out from all the other famous Bohemian characters he might have mentioned, Kepler emphasized the extraordinary nature of the circumstances in which he was writing. Libussa's prestige would not have been needed to settle a simple family spat. Civil war was brewing, Kepler warned his readers in this veiled fashion; he was telling them to look between the lines for information he felt momentarily unable to convey forthrightly.

What kind of information?

The manuscript proceeded: "It happened then on a certain night that after watching the stars and moon, I stretched out on my bed and fell sound asleep. In my sleep I seemed to be reading a book I had got from the market."

One of the most popular public markets of middle Europe of the seventeenth century was the annual book fair at Frankfurt, Germany. Kepler attended it periodically in the same spirit that takes modern scientists to the annual meetings of their professional societies. He marketed his own books, in which he reported on his varied researches, and he bought books describing the newest discoveries of other scientists.[33]

Hence the book being read in the lunar allegory could be assumed to be a book on astronomy. Since the writer of this book was Johannes Kepler, imperial mathematician to the Hapsburgs, his scientific reputation would stand behind what followed even though what followed pretended to be a dream. Surely, Kepler must have thought, no scholar worth the name could fail to pick up the message.

Of course, no one could get the message until they got a copy of the lunar geography. And the only notice of the geography's existence had been in the *Conversation with the Star Messenger*. Unfortunately, neither Galileo nor any other scientific

. . . . . .

[33] Kepler's trip to the book fair in 1609 is mentioned in Caspar, *op. cit.*, p. 177.

reader of the *Conversation* expressed interest in the new geography. The first copy of the manuscript entered private circulation through the hands of a well-to-do layman sometime during the year after Galileo was informed of the geography's existence.

According to Kepler's own subsequent account, this copy somehow got from Prague to Leipzig, and from there was taken to Tübingen by a "Baron von Volckersdorff and his tutors in morals and studies." [34] The wording suggests that the Baron was touring Europe under the chaperonage of an older scholar, as was the postgraduate study custom of that time.[35] Such a circumstance would fit logically into the procession of events that followed. Anyhow, the lunar geography began to be read privately in the year 1611 in a manner beyond Kepler's control and not always among people he would have chosen for an audience. The matter got so far out of his ken that at one point he thought the manuscript had reached England and there had inspired John Donne's devastating satire on the Catholic hierarchy, *Ignatius His Conclave*.[36]

. . . . . .

[34] Kepler's footnote 8.

[35] This probability was first pointed out by Professor Marjorie Nicolson of Columbia University.

[36] The phrasing of Kepler's footnote 8 suggests that he was merely guessing. "If I am not mistaken," this footnote begins. But Professor Nicolson contended he must have had sound footing for his suspicion. He must have known that the lunar geography had reached the hands of John Donne, the English poet who wrote the *Conclave*. However, the *Conclave* was published anonymously, and Donne's authorship of it did not become known until after Kepler's death. Furthermore, Donne himself indicated that his reference to Kepler in the *Conclave*—the stated cause of Kepler's suspicion—was prompted by a work of Kepler's other than the *Dream*. The thesis Professor Nicolson pursued so imaginatively in "Kepler, The Somnium and John Donne" (see my footnote 41) thus seems not to be supported by the record. It is a pity that her speculations could not have proved true, for there is a romantic story behind them.

Kepler had been exchanging letters with the British astronomer, Thomas Hariot, since October, 1606. The courier of those dis-

. . . . . .

patches, John Ericksen, a onetime pupil of Tycho Brahe, was in Prague, on a trip from London, in September, 1609, at the end of the summer in which Kepler had first put the lunar geography onto paper.

As Professor Nicolson suggested, astronomer Hariot was one man who could have deciphered Kepler's allegorical message. Hariot himself was a maker of telescopes. He called his instruments "perspective trunckes." Whether he invented them himself, or borrowed the idea, as Galileo did, from the Dutch lensmaker, Hans Lippershey (or was it Zacharias Janssen?), is not known. In any event, he marketed "perspective trunckes" for a dozen years and provided in his will for disposal of his tools along with the two "trunckes" through which he said he saw "Venus horned like the Moon" and "spots on the Sun." It has been claimed that Hariot preceded Galileo in these observations, which were more decisive confirmation of the Copernican theory than was the discovery of the moons of Jupiter. Surviving letters written to Hariot by one of his students, William Lower, complain about Hariot's failure to announce his own findings.

Discretion, as much as modesty, may have motivated Hariot's silence. For he was the same Thomas Hariot who wrote the history of Sir Walter Raleigh's unsuccessful colony in Virginia. At the time of the lunar geography composition in Prague, Hariot was spending a lot of his time on the river Thames, rowing to and from the Tower of London. Any public attention he might have attracted to himself almost certainly would have been used adversely by the spies surrounding Sir Walter, who for years during that period was locked up with Henry Percy, Earl of Northumberland, in connection with plots against King James I. Percy allowed Hariot to use the Earl's library and to operate an astronomical observatory on the Earl's estate in return for serving as guardian to the Earl's children and running personal messenger service to and from the Tower.

Even though the relationship which Professor Nicolson thought she saw between Kepler's lunar geography and Donne's *Conclave* cannot be confirmed, anyone who reads the two works side by side today (and it can be done quite easily; Charles Coffin's excellent selection of Donne's writings is available in a Modern Library edition) is immediately struck both by parallels and by contrasts.

Where Kepler presented himself as captive of a nightmare and tried to persuade his audience that the dream was real, Donne be-

Four years after the disappearance of the manuscript, Kepler discovered to his dismay that a garbled version of its contents had reached his home duchy of Württemberg and that the superstitious Germans who heard of it there were fascinated by one particular item: The principal character in the book Kepler pretended to be reading in his pretended dream was a young man who studied astronomy with Tycho Brahe. Since Brahe's most famous associate was known to be Kepler himself, an inevitable interpretation of the dream was that it related a personal experience. The second most important character in the allegory was the mother of the young man who studied with Brahe. In the pretended book of the pretended dream, this woman summoned spirits to guide her son to the moon.

If Kepler's mother could call spirits out of the night, what sort of person must she be?

. . . . . .

gan by signaling his readers of an intention to talk with his tongue in his cheek. Donne did not merely dream; he was

". . . in an *Extasie*, and
            My little wandring sportful Soule,
            Ghest, and Companion of my body
had liberty to wander through all places . . . in the firmament."

"All the roomes . . . of the heavens" had already been discovered and explored by Kepler and Galileo, Donne reported after his first reconnaissance. So he looked in the opposite direction "by the benefit of certaine spectacles, I know not of what making." The poet was a subtle man, but it takes no straining of the eyes to see in this passage deft recognition of the fact that although Galileo's lenses had become famous they were no match for the imaginative vision of the human mind.

Donne clearly saw how rare a being Kepler was, how lonely the path he trod in the cold light of scientifically verified reason. Would ordinary people dare to follow even an imperial mathematician into the boundless space whence Kepler beckoned his readers? That Donne thought not is evident from the care he took to put readers of the *Conclave* at ease by partitioning his heavens into "rooms" of familiar and comfortable size.

Events proved how profoundly the poet plumbed human nature. At the same time, they exposed the scientist's superhuman expectations.

The Württembergers thought she must be a witch, and Kepler's mother was therefore accused of witchcraft.

The details of her trial and confrontation with the instruments of torture will be told later in these pages. Here it is enough to say that she died of the effects of the ordeal shortly after Kepler freed her from her prison chains. Kepler himself emerged from the harrowing experience thoroughly outraged, and wrote clarifying footnotes to his allegorical geography intermittently for ten years afterward in his determination to expose the disgraceful injustice that had been done. In those notes, Kepler finally revealed the extent to which he had synthesized a lifetime of learning and applied it to the problems of extraterrestrial exploration.

Perhaps if Rudolph II's imperial mathematician had survived to supervise publication of this most advanced of all his works, the indomitable will that drove him to fame in spite of squinty eyes, twisted hands, a bowlegged body, boils, and nervous afflictions, would somehow have commanded the attention of his peers for this volume, too. However, only six pages of the manuscript were in type at the time of his sudden death. Kepler's son-in-law, Jacob Bartsch, tried to complete publication of the work but he also died before it was finished.[37] Only the extreme poverty of Kepler's widow finally drove the book into print (through the intervention of Kepler's son, Ludwig) as a family potboiler four years after the great man's death.

There should have been a good market for the book, for it appeared during the year after Pope Urban VIII sentenced Galileo to house arrest for publishing the *Dialogue Concerning the Two Chief World Systems*; popular interest in the sun-centered astronomy of Copernicus was then at a historical peak. But the record bears no indication that the royalties from the moon geography contributed greatly to the income of Kepler's survivors.

The thin volume was titled simply *Dream*, with the subtitle *Astronomy of the Moon*. The original 1634 edition was in Latin, another Latin edition was published in Volume VIII of Chris-

. . . . . .

[37] See the Dedication of the *Dream* on page 83 of this volume.

tian Frisch's collection of Kepler's works in 1870,[38] and a German paraphrase was made by Ludwig Gunther in 1898 under the title *Kepler's Traum von Mond.*

Both Frisch and Gunther understood that beneath its allegorical surface Kepler's *Dream* held a profound thesis organized by a prodigious mind.[39] But neither of the men had enough scientific background to bring the argument of the 1609 text and the subsequent footnotes together with the clarity and force required to convince readers outside Germany that a genuine scientific document lay before them.

Robert Small's *Account of the Astronomical Discoveries of Kepler* is the English classic on Kepler's work. Written in 1804, it explicitly recognized that Kepler, and not Sir Francis Bacon, was the originator of the experimental method of modern science. Of *New Astronomy* Small said, "This work . . . exhibited, even prior to the publication of Bacon's *Novum Organum,* a more perfect example than perhaps ever was given of legitimate connection between theory and experiment, of experiment suggested by theory, and of theory submitted without prejudice to the rest and decision of experiments." [40] Of the *Dream,* however, Small said nothing whatever.

In the twentieth century, Professor Marjorie Nicolson of Columbia University revived interest in the *Dream* by tracing its influence on English literature.[41] Although she pointed out the chronic neglect of the scientific content of the *Dream* footnotes by historians of science, she characterized the *Dream*

. . . . . .

[38] The *Opera Omnia.*

[39] See Frisch's preface to the *Dream* in *Opera Omnia,* VIII, and Gunther's commentary introducing the *Traum.*

[40] A new edition of this historic work was published in 1963 by the University of Wisconsin Press, with a foreword by William D. Stahlman.

[41] Marjorie Hope Nicolson, *Voyages to the Moon* (New York: The MacMillan Company, 1948), pp. 41–52; *Science and Imagination* (Ithaca, N.Y.: Great Seal Books, a division of Cornell University Press, 1956), pp. 58–79; "Kepler, the Somnium and John Donne," in *Roots of Scientific Thought,* edited by Philip P. Wiener and Aaron Noland (New York: Basic Books, 1957), pp. 306–327.

itself as "a piece of fiction," "his [Kepler's] little fictional work," thereby setting a fashion that seems to have captured all subsequent critics of the *Dream* except the German-born advance man for interplanetary space travel, Willy Ley.[42]

My own awareness of the *Dream's* existence dates from 1957, the year of Sputnik I. In the course of my reading duties as Science Editor of *Saturday Review*, I came across mention of Kepler's neglected manuscript in a popular history of astronomy written by Rudolph Thiel.[43] Assuming that all Kepler's works would be easily available in English, I walked the few city blocks that separate my office from the main reading room of the New York Public Library and sought to borrow an English copy of the *Dream*. To my astonishment, there was only the original Latin edition of 1634, the Latin volume that is part of the Frisch set of Kepler's collected works, and Gunther's German *Traum*. There was nothing in English.[44]

. . . . . .

[42] Willy Ley, *Rockets, Missiles and Space Travel* (New York: Viking Press, 1961), pp. 14–19.

[43] Rudolph Thiel, *And There Was Light* (New York: Alfred A. Knopf, 1957), pp. 128–130. The reference turned out to be inaccurate. Like others who had not carefully read Kepler's footnotes to the *Dream*, Thiel wrongly assumed that the *Dream* was written after Kepler had studied the moon through a telescope. Only the appendix was so composed. (The appendix is printed at the end of this volume and is worth reading in its own right as a species of scientific joke. I consider it a satire on the Aristotelian search for final cause, discussed further along in these pages. Having been burned once by popular misinterpretation of the *Dream*, Kepler took no chances with the appendix. "These remarks are made in sport," he said at the end of its text.) Kepler's pride in composing the *Dream* proper before a telescope became available to him is reflected in footnotes 109, 154, 165, 207, 208, 211, 213, 223.

[44] I later learned that one of Professor Nicolson's students, Joseph Keith Lane, had translated the *Dream* into English at Professor Nicolson's urging. Never published, his manuscript is on file in Columbia University's Carpenter Library under the title "The Dream; or Posthumous Work on Lunar Astronomy by Johannes Kepler" (English MA, 1947). Another English version of Kepler's manuscript which I later discovered was a translation made by

Curious, I paid library staff members by the hour to read to, and translate for, me from the Latin and German at lunchtime and at night. The more I heard the more convinced I became that I was listening to a forgotten masterpiece. How could I encourage its dissemination in English? I decided to commission a translation from my modest personal funds, and was fortunate in being able to interest Mrs. Patricia Frueh Kirkwood, wife of Professor Gordon M. Kirkwood, Chairman of the Department of Classics at Cornell University, in the assignment. With her work and my interpretation of it in hand, I asked advice of Dr. C. Doris Hellman, a protégé of the great George Sarton. She, an Adjunct Professor of the History of Science at both Pratt Institute and New York University, had translated the definitive Caspar biography of Kepler into English. After referring me to documents I had not yet become aware of, and offering some new insights and perspectives, she exhorted me to patient persistence in my determination to add the *Dream* to the regrettably small store of Kepler's writings that are available in English.

IV suppressed ~~Suballen~~ *He was extremely by authority.*

We have seen that Kepler's original intention was to circulate his *Dream* privately, among scholars familiar with Latin—but if the manuscript got any serious attention from them, no record of it survives. The *Dream* might have disappeared into the limbo of unpublished papers if a distorted notion of the manuscript had not brought about the trial of Kepler's mother as a witch.

What was there about the allegory that made it fuel for the flames of prejudice which destroyed Kepler's mother soon after she escaped burning at the stake?

Let us try to find out by returning to the *Dream* at the point where we left it, a few pages back. We find Kepler dreaming of reading a book. The main character in the book speaks out:

. . . . . .

Everett F. Bleiler and published in a science fiction anthology: *Beyond Time and Space*, selected and with an introduction by August Derleth (New York: Pelegrine and Cudahy, 1950).

My name is Duracotus. My home is Iceland, which the ancients called Thule. Because of the recent death of my mother, Fiolxhilde, I am free to write of something which I have long wanted to write about. While she lived she earnestly entreated me to remain silent. She used to say that there are many wicked folk who despise the arts and maliciously interpret everything their own dull minds cannot grasp. They fasten harmful laws onto the human race; and many, condemned by those laws, have been swallowed by the abysses of Hekla. My mother never told me my father's name, but she said he was a fisherman and that he died at the very old age of one hundred and fifty years (when I was three) after about seventy years of marriage.

In my early childhood, my mother often would lead me by the hand or lift me onto her shoulders and carry me to Hekla's lower slopes. These excursions were made especially around the time of the feast of St. John, when the sun, occupying the sky for the whole twenty-four hours, leaves no room for night. Gathering herbs there, she took them home and brewed them with elaborate ceremonies, stuffing them afterward into little goatskin sacks which she sold in the nearby harbor, to sailors on ships, as charms. In this way, she made a living.

Later, we will find another meaning hidden two layers down in this passage—a meaning that will introduce us to Kepler's scientific purpose in writing the manuscript. But first we must discover a less deeply buried meaning that will lead us to the witchcraft trial of Kepler's mother.

Who was Duracotus?

According to the *Dream* passages just quoted, he was the son of an herb doctor. And we know that Johannes Kepler's own mother, Katharina, was an herb doctor.[45] Was Duracotus really Johannes Kepler, then? The text holds another clue that may decide the answer. Was Katharina Kepler's husband an old man at the time Johannes Kepler was born?

No. The allegorical fisherman who sired a son at the age of one hundred and forty-seven years was a typical Kepler joke, lifted from something Kepler had read; he probably intended it to be laughed at and forgotten.[46] However, Katharina Kepler

. . . . .

[45] Caspar, *op. cit.*, p. 35.
[46] Kepler's footnote 11.

had plenty of reason to prefer to keep her husband's name to herself. Heinrich Kepler, Johannes' father, was a mercenary soldier who began by deserting his wife and children periodically for the excitement of battle and ended by disappearing altogether.[47]

If we read what follows in the light of the foregoing information, succeeding passages in Kepler's allegory become one of the most ingenious autobiographies ever written. For Duracotus goes on to say that "once, out of curiosity," he "cut open a pouch unbeknownst to my mother who was in the act of selling it, and the herbs and patches of embroidered cloth she had put inside it scattered all about. Angry with me for cheating her out of payment, she gave me to the captain in place of the little pouch so that she might keep the money."

Now nothing of this sort happened to Kepler as a boy. But his desperately lonely mother at one point did become a military camp follower in pursuit of her renegade husband, abandoning her son to the care of his grandparents.[48] Had young Johannes' health been sturdier, he might have spent his adult life as a plowman. Instead, his frailty and quick wit encouraged his grandfather, Burgomaster of the Free City of Weil, to obtain a fellowship for the lad to prepare for the Lutheran clergy at Tübingen University.

According to Duracotus, the allegorical ship captain who acquired the boy as a forfeit for the good luck charm, "setting out unexpectedly next day with a favorable wind, headed as if for Bergen in Norway. After several days, a north wind came up; blown off course between Norway and England, he headed for Denmark and traversed the strait, since he had a letter from a Bishop of Iceland for Tycho Brahe, the Dane, who was living on the island of Hven." By the time the boat reached shore, the boy was "violently seasick from the motion and the unusual warmth of the breeze, for I was in fact . . . only fourteen." The captain "left me and the letter with an island fisherman," to be taken to Brahe.

Since Johannes Kepler had never been sold to a shipmaster,

. . . . . .
[47] Caspar, *op. cit.*, p. 36.
[48] *Loc. cit.*

he plainly could not have undertaken the voyage just described. But Kepler had left his mother for distant parts at the age of thirteen years,[49] going first to preparatory school at Adelberg, thence to Maulbronn, finally to Tübingen, and then moving on to a teaching post at Graz; from Graz an unexpected shift in the winds of fortune (the sudden closing of the Lutheran school there by the accession of Catholics to authority[50]) drove him to the protecting arms of Brahe, the fabulously accurate Danish astronomer.

Kepler never actually visited Hven, although through association with Brahe he certainly—as Duracotus put it—"achieved an understanding of the most divine science [astronomy]." Brahe had fallen out with his Danish royal patrons, had left Hven, and was in Prague serving the Hapsburg Emperor Rudolph II as imperial mathematician when Kepler joined him in the year 1600.[51]

Upon Brahe's death in 1601, Kepler succeeded to the title of imperial mathematician under the reigning Hapsburg and inherited Brahe's observations of the oppositions of Mars during the years from 1580 to 1600. It was through mathematical analysis of those observations that Kepler finally reached the conclusion, in 1605,[52] that the orbits of the planets must be ellipses and not circles. The planet-motion laws he derived from this conclusion were published in 1609, the same year in which he wrote the *Dream*.

So, in the end, after having given misleading data on almost all the major details, Duracotus revealed himself to be an authentic mirror image of Kepler in these allusions:

"Brahe and his students passed whole nights with wonderful instruments fixed on the moon and stars. . . . The astronomical exercises pleased me greatly . . . and prepared the way for me to greater things."

The allegorical Duracotus also faithfully reflected the real

· · · · · ·

[49] Caspar, *op. cit.*, pp. 39–44.
[50] *Ibid.*, pp. 111–115.
[51] *Ibid.*, pp. 70, 118.
[52] *Ibid.*, p. 139.

Kepler when he declared, "After I had passed several years on this island [that is, with Brahe] I was seized with a desire to see my home again." For Kepler was dreadfully homesick in Prague. He tried again and again to return to Tübingen University as a faculty member. But the stiff-necked Lutherans there wouldn't have him. In this respect, the allegorical Duracotus did what Kepler longed to do but could not:

"Therefore, after obtaining my patron's approval for my departure, I left him and went to Copenhagen. There some travelers who wanted to learn the language and the region took me into their company; with them I returned to my native land five years after I had left it."

Duracotus said he found old Fiolxhilde "still alive, still engaged in the same pursuits as before." The appearance of her son "brought an end to her prolonged regret at having rashly sent . . . [him] away. . . . She said she was ready to die since her knowledge, her only possession, would [now] be left to her son and heir." Duracotus,

. . . by nature eager for knowledge . . . asked . . . what teachers she had had in a land so far removed from others. One day when there was time for conversation she told [him] . . . everything from the beginning, much as follows:

Duracotus, my son, she said, provision has been made not only for the regions you have visited, but for our land, too. For although we have cold and darkness and other inconveniences, which now at last I am aware of when I learn from you the delights of other regions, we are nonetheless well endowed with natural ability, and there are present among us very wise spirits who, finding the noise of the multitude and the excessive light of other regions irksome, seek the solace of our shadows and communicate with us as friends. Nine of these spirits are especially worthy of note. One, particularly friendly to me, most gentle and purest of all, is called forth by twenty-one characters. With his help I am transported in a moment of time to any foreign shore I choose. . . . I should like you to go with me now to a region he has talked to me about many times, for what he has told me is indeed marvelous. She called it Levania.

That allegorical conversation took place on a spring night. There was a crescent moon. Duracotus encouraged Fiolxhilde to "summon her teacher." Then she "withdrew from me to a nearby crossroads, and after crying aloud a few words in which she set forth her desire, and then performing some ceremonies, she returned, right hand outstretched, palm upward, and sat down beside me. Scarcely had we got our heads covered with our robes (as was the agreement) when there arose a hollow, indistinct voice, speaking in Icelandic. . . ."

At this point in Kepler's story, a change occurs in the identity of the narrator of the allegory, which continues to be told in the first person. The new speaker does not say who he is. Kepler merely drops the name into the middle of a page of the manuscript, as though the allegory were a play and its author were indicating the character who would deliver the next line:

THE DAEMON FROM LEVANIA

This Daemon immediately located Levania for his audience. The place was "fifty thousand German miles up in the air" and could be reached from earth only by a road that "seldom lies open." Even when the road was open, passage over it by humans was "exceedingly difficult, and made at grave risk to life." Only seasoned travelers could survive this route. The most likely candidates were "dried-up old crones . . . who since childhood have ridden over great stretches of the earth at night in tattered cloaks on goats or pitchforks."

In the footnotes which he began adding to the manuscript a decade later, Kepler explicitly blamed his mother's trial for witchcraft on the appearance of the Daemon in his lunar manuscript and the Daemon's expression of preference for the companionship of crones.[53] He seems to have been astonished by the selectivity with which illiterate hearers of his *Dream* seized a joke within an allegory and acted upon it as though the joke had been a serious statement of fact. However, given

. . . . . .

[53] Kepler's footnote 60 says, "Behold Aulis and the alliance that destroyed Troy."

hindsight, it is easy to see why events happened in such fashion. Granted that the gist of the manuscript became known to unfriendly persons, as Kepler says it did in the year 1611, we are in fact hard put to imagine how the course of affairs could have been otherwise. The autobiographical allusions of the allegory were too plain and too consistent.

Duracotus was intended to be recognized as Kepler; Fiolxhilde was intended to be recognized as Kepler's mother; Duracotus described Fiolxhilde calling the Daemon out of the night; the Daemon reported that skinny old women were the best prospects for travel to other worlds; and Kepler's mother was thin, and over seventy years of age. The one element remaining to complete a pattern for a witchcraft trial was some provocative act in the real life of Katharina Kepler, for fear of witches was rampant in Germany in those days.

Kepler, in a footnote covering this point, relates his mother to the *Dream* by indirection. He says the copy of the lunar manuscript he gave to Baron von Volckersdorff became the subject of "conversation . . . in the barbershops" among people to whom "the name of my Fiolxhilde is particularly ominous . . . by reason of their occupation." "Slanderous talk," he calls it; "taken up by foolish minds," it "became blazing rumor, fanned by ignorance and superstition." Kepler blames this barbershop gossip for bringing upon his family a "plague of six years" and requiring his presence in Württemberg for a year.

Although he used the plural, "barbershops," Kepler said the slanders came from "that house," and it has since been established, from old court records in Stuttgart, that only one barbershop was involved.[54]

Who was the barber?

The name can be found in Dr. Hellman's translation of Max Caspar's biography of Kepler by anyone who undertakes the search from the proper perspective. Caspar describes the writing of the *Dream* by Kepler and the witchcraft trial of Kepler's mother in reverse order from the real sequence of the two

. . . . . .

[54] Caspar, *op. cit.*, pp. 243–245

events. The trial became a matter of public record, in the court at Stuttgart, during Kepler's lifetime. The *Dream* reached public print after Kepler's death. Caspar therefore noted the trial first[55] and the appearance of the book second.[56] By turning Caspar's text around and putting his account of the *Dream* where it belongs chronologically, ahead of his account of Katharina's court troubles, it is seen that the barber mixed up in the witchcraft case was Urban Kräutlin,[57] who served the princeling brothers of Duke John Friedrich of Württemberg. Like all other barbers of that day, Urban Kräutlin was, as Kepler's reference to "their occupation" indicates, also a minor surgeon, knowledgeable in the prescription of natural drugs.

How the barber Kräutlin heard of Kepler's *Dream* can only be surmised. But Kepler gives us some clue in the footnote in which he reports giving a copy of the original manuscript to "Baron von Volckersdorff and his tutors in morals and studies." The Baron was probably making the grand tour of Europe at the time. As imperial mathematician to the reigning Hapsburg, Kepler would have been a desirable personage for the young Baron to meet. And Kepler, chronically uncertain about collecting the sizable amounts of back pay due to him, might logically have gone out of his way to please a rising young member of the governing class. If Kepler had talked to the Baron about the unpublished *Dream*, the Baron certainly would have been flattered by the confidence and might very well have asked to read the manuscript. Being an unusually generous man, Kepler could have made a typical gesture by arranging to lend the Baron the first copy. In that event, the young visitor likely would have talked about the *Dream* wherever he went, and might even have asked members of the faculty at Tübingen, Kepler's alma mater, to interpret difficult passages for him.

This reconstruction of events is necessarily speculative. But it fits Kepler's statement that the lunar allegory stirred up

. . . . . .

[55] Caspar, *op. cit.*, pp. 240–258.
[56] *Ibid.*, pp. 351–353.
[57] *Ibid.*, p. 243.

trouble for his mother. The exact, or even approximate, time of the start of this trouble is not clear. But according to the narrative line recited by Caspar,[58] it must have been before the barber Kräutlin accompanied Duke John Friedrich's brother, Prince Achilles, on a hunting trip into the neighborhood of Leonberg, Katharina Kepler's home town.[59] For it was during this hunt that the barber stopped in Leonberg to visit his sister, Ursula, wife of Jacob Reinbold, the town glazier.

Ursula had been a crony of Katharina's for many years, but the friendship had recently soured. The reason was that Ursula, unknown to her brother the barber, had become pregnant through relationship with a man other than her husband and had submitted to abortion to escape scandal. She had confided in Katharina, and Katharina had told Kepler's younger brother, Christoph, a tinsmith's apprentice who exercised his lungs as drill sergeant of the Leonberg town guard. Christoph also sounded off in public about Ursula's indiscretion and his mother foolishly gave confirmation. To cover up the truth, Ursula attributed her debility to an evil spell and blamed Katharina Kepler for casting the spell.

The barber first heard his sister's tale in the midst of a drinking bout with her and her husband and the Leonberg Magistrate, Luther Einhorn. In a sodden brain the accusation would have fitted perfectly with the summoning of the allegorical Daemon to guide Johannes Kepler to the moon. Magistrate Einhorn already had a grudge against the Kepler family. He had wooed Johannes' sister, Margarete, a few years before, only to be humiliated in the eyes of the whole town by her marriage to a milksop Lutheran preacher, Georg Binder.

While the night's libations were still in charge of their faculties, the vengeful Ursula, her credulous husband, her vindictive brother the barber, and the jealous judge summoned Katharina Kepler to the Leonberg town hall. There the princes' barber

. . . . . .

[58] *Ibid.*, pp. 240–256.
[59] *Ibid.*, p. 244.

melodramatically drew his sword, poised its point on Katharina's breast, and threatened death on the spot unless the old woman would say the magic words to make his sister well again.

If Katharina had possessed common sense to match her temper, she might have realized, by her escape from this ordeal unharmed after a simple denial of magical powers, that no participant in the brawl would have welcomed even a clear memory of it next day. But, being a confirmed brawler in her own right, Katharina made the affair public by suing her estranged friend Ursula for slander.

Katharina's case was hopeless from the start. The livelihood of the princes' barber would hardly survive official recording of his threat against the life of an innocent woman. Nor could the Magistrate afford to be formally associated with the drumhead court procedure he had in fact condoned. And Ursula was safe as long as she kept stirring the witch's brew against Katharina.

Ursula's assignment was not difficult. The atmosphere she lived in did half the job. Caspar points out that Germany was engulfed in a mass madness the like of which would not be felt again until Adolf Hitler appeared three and a quarter centuries later. A fear of witches was everywhere. Catholics and Protestants alike, people of noble birth and good education along with ignorant peasantry, fell prey to a prevalence of spells. Accidents to men and animals were traced to instructions passed up from the nether regions by word of mouth. Innocent acts became vicious conspiracies. Anything unusual seemed mysterious, and even the usual was twisted into something devilish and bizarre. Kepler's native region of Weil der Stadt, populated by no more than a few hundred families, condemned thirty-eight women to death for exercise of extraordinary powers during the years 1615 to 1629. In the town of Leonberg alone six were burned as witches within a few months at the end of 1615.[60]

. . . . . .

[60] Caspar, op. cit., p. 241. Students of historical parallels will observe that this period was also a time of ordeal for Galileo across the Alps in Italy. He traveled from his home in Florence to Rome in 1615 to defend himself against an attack that had been made on him from a Dominican pulpit the year before. The trip failed of its

The drunken confrontation of Katharina Kepler in the Leon-berg town hall took place in August, 1615, at the start of that era of terror. Tongues began to wag in ominous tune. Hadn't old Katharina been reared by a female kinswoman executed for witchery?

The affirmative answer would explain many strange events, such as Katharina's request to the sexton of the Eltingen churchyard for the skull from her father's grave, and her with-drawal of the request when she was told she'd first have to get approval from the local authorities. What was the business about her wanting to get the skull silvered to serve her son Johannes as a goblet? What sort of ceremonial drink would such a goblet grace? And the tale that Katharina's epileptic son, Heinrich, told about his mother riding a calf to death and then feeding Heinrich a veal chop from the unlucky animal— surely the woman knew quieter ways of killing a beef!

The invisible net of suspicion around the imperial mathema-tician's meddlesome parent spun steadily tighter as she gadded around the streets in the restless nosiness of her declining years. Beutelspacher, the lame schoolmaster, vividly recalled that his disability had followed a drink he had taken from a tin cup at Katharina's house one night ten years before while he was read-ing one of Johannes' letters to her. Another neighbor, Bastian Meyers, remembered that his wife had withered and died after a draft from the same tin cup.

The wife of the butcher, Christoph Frick, swore that one of her husband's thighs was stabbed with pain when Katharina passed. Katharina walked on, but the pain remained until Frick mumbled to himself in church, "Katharina, help me, for God's sake!" He hadn't then noticed Katharina's presence at the service, and she sat too far away from him to hear what he said. Nevertheless, she turned and looked in the butcher's di-rection after he uttered the words. Thereupon the pain "blew away."

Daniel Schmid, the tailor, held Katharina responsible for the
. . . . . .
purpose, and in 1616 the Vatican condemned the Copernican the-ory, ordering Galileo to abandon his advocacy of it. See Drake, *op. cit.*, pp. 284–285.

deaths of his two children; time after time she had entered his house uninvited and had said incantations over their cradles; the dreadful part of it was that she had seemed to be blessing them. Katharina had given Schmid's wife some words to speak in the graveyard in the light of the full moon, and Schmid's wife had spoken the words, but the children died anyhow. The words obviously weren't the right ones, and Katharina knew it.

Cattle tethered in barns in the town stamped in their stalls at night and went mad. No one could think of any other reason for this than Katharina's spell. There was no escaping the woman. Why, she could even walk into houses when all the windows were shut and the doors were locked!

As one after another of her fellow townsfolk discovered Katharina's involvement in unhappy aspects of their lives, Ursula Reinbold proved the old woman's complicity in her own malaise with the help of Jeorg Haller, a day laborer's wife. Mrs. Haller's evidence consisted of measurements of Ursula's head.

Katharina unfortunately accommodated her detractors by continuing her restless rounds of unwanted advice. Worse still, she kept urging her homemade remedies upon everyone for every sort of indisposition, always from the now-notorious tin cup.

The disputatious crone's position deteriorated steadily from August until December. Only then did her son Christoph, her daughter Margarete, and Margarete's husband, the weak-spined Pastor Binder—all three of whom had joined in putting the fat in the fire with the suit for slander—decide that it might be wise to seek advice and help from Johannes.

Margarete first wrote to her brother in the autumn of 1615. He received the letter on December 29, 1615, in the sleepy Austrian town of Linz.[61] He had moved to Linz from Prague to escape a long series of personal tragedies that had begun in the year when Baron von Volckersdorff got the first copy of the *Dream*.

In the earliest of these misfortunes, Kepler's royal patron, the

. . . . .

[61] Caspar, *op. cit.*, p. 244.

Emperor Rudolph, had been forced to abdicate[62]—an eventuality the imperial mathematician had foreshadowed in the opening paragraph of the *Dream*. The coup was accomplished by a movement of troops under Rudolph's brother, the Archduke Matthias. With the soldiers came a smallpox epidemic which attacked, among others, Kepler's favorite son, Friedrich, "a hyacinth of the morning in the first day of spring." The boy died and his mother, Barbara, Johannes' wife, was first deranged by grief and then herself carried off by the pox.[63] Nonetheless, Kepler remained in the service of the deposed Rudolph until Rudolph died in January, 1612.[64]

Matthias continued Kepler's appointment as imperial mathematician but did not consult him often and agreed to allow him to move to Linz in Austria and enjoy the additional salary of provincial mathematician there.[65] A certain amount of political liberality may have been entailed in this act of the new Emperor; my own belief is that a greater influence was the recognition by the Hapsburg bankers, the Fugger spice- and drug-trading family, that the planetary timetables Kepler had promised to derive from his new laws of orbital motion would be of incalculable value to global shipping.

Kepler had barely settled down in Linz with his second wife, Susanna, before new woes beset him. The Lutheran pastor in the town, who had known the Kepler family in Württemberg and disapproved of Kepler's broad-mindedness, refused to allow the devout scientist to participate in the Holy Communion.[66]

It was thus a spiritually bruised, emotionally bleeding Kepler who opened the letter from his sister Margarete and read "with unutterable distress" the miserable news about his mother. He felt his heart "nearly . . . burst in my body." In spite of the disgust he must have felt over the family's bungling, in spite of the resentment that would have been a normal and natural

. . . . . .

[62] *Ibid.*, p. 203.
[63] *Ibid.*, p. 206.
[64] *Ibid.*, p. 208.
[65] *Ibid.*, p. 211.
[66] *Ibid.*, pp. 213–214.

reaction to their disregard of his own position, Kepler responded sympathetically. Perhaps he appreciated that if his mother had lacked the meddlesome quality that kept her in constant trouble, he might not have inherited the curiosity that made him a genius of observation and synthesis. Anyhow, he quietly added her tragedy to the burden already bowing his shoulders.

On January 2, 1616, three days after receiving Margarete's letter, he wrote to the Württemberg officials demanding documentation of the charges against his mother, thereby discovering that he himself also had been accused "of forbidden arts." [67] Since there was nothing against him except the *Dream*, he was never again implicated, and his call for evidence might even have extricated his mother if she had been a less contentious person. But she persisted in trying to bring her slander suit against Ursula Reinbold to trial.

The princes' barber and the Leonberg Magistrate, of course, continued to use their positions to delay a court hearing of Katharina's complaints. But there are practical limits to the effectiveness of such political chicanery. Finally, the two men realized that to save themselves they had to weave the superstitious gossip about Katharina's ability to cast spells into a convincing case of witchcraft against her. As a witch, she could not be slandered even by the most obvious falsehoods.

The pattern of the web of conspiracy was designed around a child: the twelve-year-old daughter of Haller, the laborer whose wife had measured Ursula Reinbold's head.

The girl had been carrying bricks to a kiln outside the town when Katharina Kepler, on her busybody rounds of the countryside, brushed against the child's arm. Pain set in at once, growing every hour until the girl could move neither hand nor fingers. The girl's mother rushed at Katharina with a knife, shrieking for restoration of the muscles of her daughter's arm. The princes' barber reappeared on the scene through some magic peculiar to him, and the Magistrate, after an examination of the girl, pronounced her arm to be "in a witch's grip." Katharina then confessed guilt in the eyes of everybody but her son

. . . . . .

[67] Caspar, *op. cit.*, p. 245.

Johannes (even Christoph Kepler and the spouses Binder had told the Magistrate they would disown her if Katharina were found guilty of witchcraft) by offering the Magistrate a silver chalice if he would stop the proceedings.

This compromising gesture was made late in the year 1616. Only one escape path remained open for Katharina Kepler after that. It led to Johannes in Linz. With Christoph as an escort, she reached Johannes' home on December 13, 1616. Behind her in Leonberg a warrant was issued for her arrest.

As long as the old woman stayed with Johannes she was safe. But Katharina did not like Austria. In October, 1617, she slipped away from her son and returned to Margarete in the university town of Heumaden. The imperial mathematician patiently traced her route of travel and followed her, reading on the way a book written by Galileo's father, *Dialogue on Ancient and Modern Music*. Many Germans were taking refuge in music in those unhappy times, among them the Bach family in Brandenburg. To Kepler, the stars were a form of music, and despite the excruciating tension of his journey, he drew from the Galileo book inspiration for his own next work, *Harmony of the World*, in which he stated his third law of planetary motion, that the square of the time of one revolution of a planet in its orbit is proportional to the cube of that planet's mean distance from the sun.[68]

Kepler remained by his mother's side for two months, writing various petitions on her behalf. His only success was to obtain a permit to take her back with him to Linz. But she refused to go.

For the next two years—the opening years of the Thirty Years' War—the warrant for her arrest lay unserved in the Leonberg Magistrate's office. During that period, forty-nine "points of disgrace" were drawn up, including a new charge that Katharina Kepler had tried to apprentice into witchcraft a hunter's daughter named Barbara. In the spring of 1620 Kepler wrote from Linz to the Duke of Württemberg asking the right to answer for his mother and clear the family name. The response

· · · · · ·

[68] *Ibid.*, p. 286.

was an order, issued July 24, 1620, that Katharina Kepler be arrested on sight. And on August 7 of that same year, at the age of seventy-four, she was seized in Margarete's house in the middle of the night and carried out bodily in a linen chest to avoid public outcry. Next day, the Magistrate heard her denial of being a witch, and committed her to prison for a second interrogation before putting her to the torture.

When word came from Margarete about the way Katharina had been taken from the house in a box in the darkness, Kepler uprooted his entire household from Linz to go to his mother's help. The war was just then beginning to engulf the city; not daring to leave his wife and children behind, he moved them to the safety of Regensburg and continued alone to Leonberg to defend the old woman's life.[69] Upon reaching Leonberg he discovered that his mother had been moved to the town gate of nearby Guglingen because her confinement in Leonberg embarrassed her drill-sergeant son, Christoph.[70] Johannes found her in chains in a cold stone room.

The imperial mathematician's unexpected appearance had the immediate effect of delaying the trial for six weeks. After retaining a boyhood chum, Christoph Besold, as legal counsel, the great scientist went to work scientifically on the case himself, hunting natural causes for the supposedly supernatural events with which his mother was charged. From Guglingen to Tübingen, from Tübingen to Stuttgart, from Stuttgart to Leonberg, from Leonberg to Guglingen he went, collecting evidence.[71] Caspar tells some of the findings.

Kepler confirmed that Ursula Reinbold had had an abortion.

. . . . . .

[69] Caspar, *op. cit.*, p. 252.

[70] *Ibid.*, pp. 252–253.

[71] Kepler's political shrewdness asserted itself again here. Although his footnotes to the *Dream* make amply clear his disbelief in the existence of witches, he did not take that position in the face of the hostile public opinion surrounding his mother's trial. He contended only that *she* was not a witch. In various petitions to the Duke of Württemberg, he stated that he realized she was a troublemaker, "garrulous," "fickle and even malicious." Caspar, *op. cit.*, pp. 240–241. Also see Baumgardt, *op. cit.*, p. 160.

Beutelspacher, the schoolmaster, had been lamed while jump-ing a ditch. Christoph Frick, the butcher, was afflicted with lumbago. The Haller girl had carried too many bricks at a time and had numbed her fingers; the blood in her arm had returned to normal circulation even before the Magistrate diagnosed her as a victim of "witch's grip."

Such painstaking documentation could only irritate Katha-rina's persecutors. They insisted that she be tortured for her stubborn refusal to confess. The Duke of Württemberg turned to the legal faculty at Tübingen University, which (no doubt affected by the influence of Besold, a distinguished member of the faculty) decided that the accused should be taken to the torture chamber and confronted with the diabolical instru-ments there—but not actually put on the rack. On September 28, 1621, this sentence was carried out, and Katharina collapsed in the midst of the ordeal. Then, after fourteen months of imprisonment, she was freed on October 4, 1621, with a warning not to return to Leonberg lest she be lynched.

Kepler now returned to Regensburg for his wife and family, and took them back to Linz. Since he hadn't explained his abrupt departure to anyone, not even to his scientific assistant, the town gossips had worked overtime during his year-long ab-sence to manufacture appropriate accountings. Some said he had gone to England, where King James was known to be friendly to him. Some said a price had been placed on his head in the course of the continued wrangling within the Hapsburg family. The latter rumor was understandable, for on June 21, 1621, twenty-seven Protestant leaders had been executed in the public square in Prague, among them men who had been Kepler's associates at Emperor Rudolph's court. The imperial mathematician's absence from Linz that summer may have been one of the luckiest happenings of his life. On the other hand, if my earlier supposition about Fugger influence is correct, those promised planetary tables were as important to the Fuggers as ever, and by the end of 1621 Kepler was not only reappointed to his old official post but exempted from a Catho-lic ban on "preachers and non-Catholic schoolmasters." [72]

. . . . . .

[72] Caspar, *op. cit.*, pp. 257–258.

A less kind fortune attended Kepler's mother. Weakened by the shock of her imprisonment and trial, she died on April 13, 1622. Later in that year, the lawyer Besold, who had defended her in court, wrote Kepler a consoling letter. Through this communication Kepler also learned that his exposure of the abuses in her arraignment had brought from the Duke of Württemberg a decree forbidding further witchcraft trials without prior sanction of the supreme court in Stuttgart.

"While neither your name nor that of your mother is mentioned in the edict, everyone knows what is at the bottom of it," Besold wrote. "You have rendered an inestimable service to the whole world, and someday your name will be blessed for it." [73]

Faithful Besold! It was the same Christoph who, as a schoolmate at Tübingen University in 1593, had offered to argue Kepler's unpopular theses about the moon in open debate. Almost thirty years had passed, and the public argument had not yet taken place.

It was not destined to take place. Kepler worked on the scientific documentation of the lunar geography sporadically for the rest of his life, but other projects kept pressing the footnotes aside. First a handbook of Copernican astronomy had to be written;[74] otherwise, the scholarly community would not understand how to revise Copernicus' thinking to arrive at a sun-centered planetary system that would work. After the handbook was done, the *Rudolphine Tables* remained to be finished[75]—the planetary position predictions that would prove the validity of Kepler's own theory of elliptical orbits. In Linz, the war delayed him; troops were barracked in his house.[76] Next, his library was sealed against him by the Catholics as the uproar over Galileo in Italy reverberated across the Alps. Seeking

• • • • • •

[73] Olaf Saile, *Troubador of the Stars*, translated by James A. Gelston (New York: Oskar Piest, 1940), p. 303.

[74] Caspar, *op. cit.*, pp. 293–300.

[75] *Ibid.*, pp. 308–318.

[76] *Ibid.*, p. 319. Worse still, his printing press was destroyed in a fire set by angry peasants.

friendlier surroundings, Kepler moved his family to Ulm, encountered other troubles there, and moved again to Sagan. In Sagan, the moon geography footnotes at last were completed.[77] A few pages of the manuscript were in type when Kepler left on a trip to Regensburg in search of funds. Death found him there a few days before a lunar eclipse.

V

During the decade between the end of his mother's trial and his own death, Kepler wrote 223 footnotes[78] to the *Dream*. They filled several times as many pages as the text to which they were appended. In them Kepler exposed another layer of meaning within the allegory—the serious scientific core of the manuscript: the geography of the moon. To understand it, we must return to the book Kepler pretended to dream he was reading, and look beneath the surface meaning that got his mother into trouble. Here again are the words:

My name is Duracotus. My country is Iceland, which the ancients called Thule. Because of the recent death of my mother, Fiolxhilde, I am free to write of something which I have long wanted to write about. While she lived, she earnestly entreated me to remain silent. She used to say that there are many wicked folk who despise the arts, and maliciously interpret everything their own dull minds cannot grasp. They fasten harmful laws onto the human race; and many, condemned by those laws, have been swallowed by the abysses of Hekla. My mother never told me my father's name. . . .

On consulting Kepler's footnotes, we now discover in the allegory a totally new set of characters.

Duracotus represents Science.[79]

. . . . . .

[77] *Ibid.*, pp. 351–353.

[78] The footnotes alone bear a date. In the original edition of the *Dream* they were grouped at the end of the text under the heading, "Johannes Kepler's Astronomical Dream Notes, written successively between the years 1620–1630." A previously quoted letter of Kepler's, written in 1623, indicates that the footnotes were actually begun in 1621 rather than in 1620.

[79] Kepler's footnote 3.

Fiolxhilde symbolizes the Ignorance from which Science springs.[80]

The husband of Fiolxhilde and father of Duracotus is nameless because the sire who brings Science forth from Ignorance[81] is Reason, and Reason is unknown to Ignorance.

Fiolxhilde is a herb doctor because primitive medicine arose from the earliest stirrings of Science within Ignorance.[82]

Given this new context, how should we interpret Duracotus' remark that, "because of the recent death of my mother, Fiolxhilde, I am free to write of something which I have long wanted to write about. While she lived, she earnestly entreated me to remain silent"? What did Fiolxhilde mean when she told Duracotus, "There are many wicked folk who . . . maliciously interpret everything their own dull minds cannot grasp" and "fasten onto the human race . . . harmful laws" under which many are condemned to be "swallowed by the abysses of Hekla"?

Because Kepler's footnote on this point[83] begins by saying that it was natural for a son to await his mother's passing rather than to expose her secrets while she lived, this reference has been interpreted as applying to Kepler's own mother's witchcraft trial and subsequent death. But such an interpretation cannot be supported. The passage is part of the original text of the allegory, which was never revised but only elucidated by means of the footnotes.[84] At the time the original manuscript was written, Kepler's mother was still very much alive. Indeed, she had not yet become involved in the neighborhood squabbling that this manuscript would blow up into tragic proportions.

. . . . . .

[80] Kepler's footnote 3.

[81] Kepler's footnote 10.

[82] Kepler's footnote 14.

[83] Kepler's footnote 3.

[84] In the previously cited 1623 letter to Bernegger, which Christian Frisch quoted in his preface to the Latin text of the *Dream* in *Opera Omnia*, Vol. VIII, Kepler wrote: "I began to revise the Lunar Geography, or rather to elucidate it with notes, two years ago."

Farther along, in the appropriate footnote, we learn from Kepler himself what Duracotus' newfound freedom was. Duracotus, be it remembered, was Science. By the year 1609, Kepler tells us, he had come to believe that "the general opposition of mankind" to Copernicus' science of sun-centered astronomy "was . . . sufficiently extinct." [85] That is, the allegorical death Duracotus intended to celebrate was the passing from university teaching of prejudice against the Copernican idea.

By "the wicked folk who maliciously interpret everything their own dull minds cannot grasp," Kepler meant the Lutheran theologians who had made his life miserable for many years before 1609.[86] And the prison assigned to those condemned by the churchmen—"the abysses of Hekla"—was the firepit of the volcano within Iceland's Mount Hekla, which medieval folktales linked to purgatory.[87]

In short, although astronomers who accepted the Copernican doctrine had previously been consigned to hell by Luther and Melanchthon, Kepler, at the time he wrote the original text of the lunar geography, thought it was safe for science to walk around hell's vestibule without being snatched inside. Apparently he was misled by Wackher's enthusiasm in the summer talks of 1609 into supposing that the hostile Tübingen attitudes of 1593 and 1597 would be reversed if the true meaning of the geography of the moon were veiled in allegory and if the manuscript were circulated discreetly.

Kepler's footnotes to the geography reveal how the manuscript was constructed to make the strongest appeal to Aristotelian prejudice without abandoning genuine scientific principles. That science always came first is established by Kepler's repeated reminders of the Tübingen theses of 1593.

Except that they treated the moon as a body comparable to the earth, the content of those theses is known only in the broadest sense. They were never published, either in their

. . . . .

[85] Kepler's footnote 3.

[86] Kepler's footnotes 3, 6, and 7.

[87] Kepler's footnotes 2, 8, and 9. Hekla had undergone one of its periodic eruptions only twelve years before the writing of the *Dream*.

original form or as excerpts.[88] Kepler indicates that they were preserved through a long stretch of his lifetime,[89] but no surviving record tells what happened to them after he finished the footnotes to the lunar geography. We can be sure only that they were made up of deductions and extrapolations of an astronomical nature, and that their nature displeased the Lutheran church authorities at Tübingen.

Being a person of enormous determination, Kepler refused to accept Tübingen debating-referee Vitus Muller's 1593 rejection of the theses as final; he held onto his manuscript against the time when it might be possible to refurbish the ideas. The problem, of course, was to lessen the fright reaction the theses provoked among conservative scholars. Apparently, a possible solution was to discover precedents for his thinking in classic Greek literature that were acceptable to Aristotelians.

Learning to read Greek was among Kepler's academic requirements in any event. To facilitate the process, he bought a copy of a then-current German translation of an old Greek account of a voyage to the moon: Lucian's *A True Story*. The translator was a son of the noted satirist George Rollenhagen. By reading the German alongside Lucian's original Greek text, Kepler mastered Greek.[90]

Lucian didn't make the slightest bit of sense scientifically, for he got to the moon in a whirlwind. But by enduring the

. . . . . .

[88] *Opera Omnia*, I, 188 n. 5.

[89] Kepler's footnote 2, written sometime after 1621, says: "I still have an old paper written by your hand, most illustrious D. Christoph Besold, when in 1593. . . ." And in footnote 43 Kepler refers again to Besold and the 1593 episode, saying "I still find these erasures in my first copy."

[90] Kepler's footnote 2. Indeed, the student Kepler mastered Greek so thoroughly that his language professor at Tübingen, the famous Hellenist, Martin Crusius, tried to obtain his collaboration in a commentary on Homer. See Caspar, *op. cit.*, p. 45. Still later, after reading Plutarch, Kepler boldly reconstructed missing pieces of that great essayist's *The Face on the Moon*. Kepler's Latin version of *The Face* was included in the original edition of the *Dream*; it is not included in Mrs. Kirkwood's translation here.

account of Lucian's hilarious discoveries on the moon, Kepler emerged with enough linguistic confidence to tackle the original Greek text of Plutarch's *The Face on the Moon*.[91]

Here was a splendid short course in ancient Greek scientific study of the moon, offered as a popular symposium. Aristarchos of Samos and Hipparchus were quoted extensively. A millenium and a half before Kepler read their words, Greek astronomers had concluded from visual observation that the moon "practically grazes the earth" compared to its vast distance from the stars, that the moon's orbit was such as to carry it sometimes nearer and sometimes farther away from earth, that the light of the moon was reflected light of the sun, and that the phases of the moon were owing to the moon's passage in and out of the shadow thrown by the earth as the earth moved around the sun.[92]

The main point of Plutarch's symposium was that, in order to

. . . . . .

[91] Kepler's road to Plutarch was purely scientific. *Dream* footnote 2 tells us that Kepler first learned about Plutarch's *Face* in reading Erasmus Reinhold's *Commentary on Peurbach's Planetary Theories*. Georg Peurbach was Europe's leading astronomer of the fifteenth century, the first to use trigonometry. With the help of a gifted pupil, Johann Muller (nicknamed Regiomontanus), he designed and built improved models of the ancient oriental astrolabes and with those instruments determined that the planets did not actually occupy the positions given for them in astronomical tables based on Ptolemaic theory. Peurbach thus broke ground for Copernicus. Ephemerides later worked out by Regiomontanus guided the voyages of Christopher Columbus. Copernicus failed to follow up the pioneering of Peurbach and Regiomontanus, however, leaving his heliocentric theory weak for lack of supporting observations. A few years after Copernicus died, Erasmus Reinhold published a wholly new set of planetary tables (the *Prutenic*), but these were still not accurate and it remained for Tycho Brahe to appreciate and fill the need for a solid body of meticulous observation of planet positions over a long period of time.

[92] I have made use of Harold Cherniss' English translation of *The Face on the Moon* in The Loeb Classical Library edition of Plutarch's *Moralia* (Cambridge, Mass.: Harvard University Press, 1957), XII, 2–223.

reflect sunlight, the moon had to be a solid body like the earth, and that the shaded areas visible on the lighted face of the moon must therefore be shadows cast by lunar mountains or depressions deep enough to elude sunlight falling on otherwise level terrain.[93]

The astronomical arguments advanced in support of this proposition were drawn from the highest scientific authorities. Writing in the spirit of his age, Plutarch melded astronomy and metaphysics and attempted to give an answer to the metaphysical question of final cause. According to the teachings of the Platonists and Aristotelians, nothing existed without cause. Therefore, the moon could be like earth only if the moon were intended to be, as earth was, a habitation for living creatures.[94]

The latter part of Plutarch's symposium was accordingly given over to discussion of the moon as a possible home for life—if not human, then some other species better suited to the lunar environment. It was noted that although life on the moon might sound preposterous, it had no greater theoretical obstacle than the idea of life in salt water might have for air-breathers ignorant of the realities of terrestrial seas. Indeed, one contributor to the discussion suggested that moon-dwellers might look upon earth's residents as the dregs of creation.[95] As a final touch, Plutarch reported the story of a mythical traveler who had sailed five days northwest of Britain to a group of islands whose inhabitants were familiar with routes of passage to the moon.[96]

When Kepler read Plutarch's reference to islands far west of the then-known world, he remembered some folktales that had been bound into the same book covers with the Rollenhagen translation of Lucian's *A True Story*.[97] The heroes of that medieval lore had been St. Brendan—a mythical Irish friar who mistook a whale for an island in the Atlantic and thereby

. . . . . .

[93] From the Cherniss translation.
[94] *Ibid.*
[95] *Ibid.*
[96] *Ibid.*
[97] Kepler's footnote 2.

baffled mappers of the western ocean for several centuries—
and the equally mythical St. Patrick, who set up purgatory
inside the volcano cone of Mount Hekla in Iceland.

The tale of St. Patrick sent Kepler's agile memory back again
from Lucian to Plutarch. For in Plutarch's symposium the route
followed by voyagers to the moon had been heavy with traffic
of the souls of the unborn and the dead, wailing back and
forth between the earth and the moon in the shadow of the
earth.

Here was the classical background Kepler needed as a setting for his Tübingen theses of 1593. Although he read the
Plutarch symposium in 1595,[98] the structure of the lunar geography did not take shape in his mind until after the solar eclipse
of October 12, 1605.[99] Then he saw that an eclipse would give
the illusion of solving certain problems inherent in any real
voyage to the moon.

The Emperor Rudolph unwittingly precipitated the actual
writing of the lunar geography by asking Kepler about the pattern of light and shade on the surface of the full moon. Rudolph
thought the shapes of the land masses of earth were reflected by
the moon, as they would be by bodies of water. The Emperor
believed he could make out the mirrored shoreline of the Italian
peninsula and nearby islands.[100] Did Kepler agree?

Rudolph's imperial mathematician emphatically did not
agree. He felt confident that the light and shadows on the
moon signified differences of elevation of the lunar landscape,
as the Greek scientists had argued in Plutarch's symposium.
And Plutarch was not the only authority to be cited. Tübingen's
astronomy professor Maestlin had taken a similar position in
talks with Kepler two decades past. Kepler discussed the Emperor's confusion with Wackher in the summer of 1609;
Wackher urged Kepler to make his ideas known, and "to
please Wackher" the geography of the moon was finally put
onto paper.

. . . . . .

[98] *Loc. cit.*
[99] Kepler's footnote 49.
[100] *Opera Omnia,* II, 290.

Whereas Plutarch's *The Face on the Moon* began with physics and ended with metaphysics, Kepler reversed the pattern in
anticipation of his critics at Tübingen. He appealed to the
Aristotelian need for an answer to the question of final cause by
implying that the moon was habitable at least by the disem-
bodied souls of men if not by living men. Souls in those days
were far more important than warm bodies, a fact Kepler recognized by summoning a representative spirit from the moon to
Iceland, the site of St. Patrick's earthly purgatory.

Aside from its place in a context of supernatural mumbo
jumbo, Iceland was a historically authentic jumping-off place
for an adventure in spherical astronomy. Because of his personal history, Kepler was peculiarly equipped to appreciate
the relationship. He had learned much lore of the far north
from Tycho Brahe, who numbered a bishop of Iceland among
his astronomy pupils at Hven. Furthermore, the traditional
occupations of Kepler's own family were weaving and fur-
dressing,[101] and their home province of Swabia by the Black
Forest in Germany was also headquarters for the Fugger
textile, dye, and spice-trading cartel.[102]

From a modest beginning as weaving-guild agents for the
powerful merchants of Venice, the Fuggers had risen to control
European commerce with the Orient through the Venetian
monopoly on Mediterranean trade from the Egyptian port of
Alexandria, Abyssinia, and points east of Africa.[103] It was to
break this monopoly that Portugal's Prince Henry the Navi-
gator spent years searching for a southern route around the
African continent. And it was from Florence, the traditional
trading rival of Venice, that the geographer Paul Toscanelli
mailed the letters that finally persuaded Christopher Columbus
to sail west to reach the East Indies.[104]

. . . . . .

[101] Caspar, *op. cit.*, pp. 29, 32.

[102] Charles McKew Parr, *So Noble a Captain* (New York:
Thomas Y. Crowell, 1953), p. 156.

[103] *Loc. cit.*

[104] *Ferdinand Columbus, the Life of the Admiral Christopher
Columbus by His Son*, translated and annotated by Benjamin Keen
(New Brunswick, N.J.: Rutgers University Press, 1959), pp. 19–22;

The Fuggers were too shrewd to ignore the possibility of cheaper access to the Oriental luxuries on which the family wealth was founded. Fugger money backed the globe-girdling voyage of Ferdinand Magellan[105] and John Cabot's search for a northwest passage around the American continent that blocked Columbus' route to Cathay.[106] The Fuggers also financed much-less-well-remembered Dutch expeditions that hunted a northeast passage to China through the Arctic Ocean.[107]

In several of the footnotes to the lunar geography Kepler refers to the Dutch explorers,[108] at one point emphasizing that they found conditions in the far north precisely as astronomers in the midst of Europe had known them to be—that is, that the tilt of the earth as it moved around the sun kept the sun below the arctic horizon continuously during winter and above the horizon continuously during summer. The Dutch had wintered in Novaya Zemlya[109] (in Russian: new-found land) and had gone on from there to discover Spitsbergen, which some mistook for Greenland, a colony of Iceland. A century before, a visit to Iceland had been part of Columbus' preparation for his daring westward plunge.[110] And sea trading between Bristol and Iceland had long preceded the voyages of John Cabot and his son Sebastian along the North

. . . . . .

cited by Samuel Eliot Morison in *Admiral of the Ocean Sea* (Boston: Little-Brown, 1942), pp. 85–86.

[105] Parr, *op. cit.*, p. 171.

[106] *Ibid.*, p. 174.

[107] *Ibid.*, pp. 53, 173.

[108] Kepler's footnotes 2, 39, 41.

[109] Kepler's footnote 2. Also see Gerrit de Veer, *The Three Voyages of William Barents to the Arctic Regions*, 2d edition, with an introduction by Lt. Koolemans Beynen, R.N.N. (originally published in London: the Hakluyt Society, 1856; reprinted by permission in New York: Burt Franklin, 1964).

[110] J. A. Williamson, *The Cabot Voyages and Bristol Discovery Under Henry VII* (Cambridge, England: Cambridge University Press for the Hakluyt Society, 1961), p. 13; Morison, *op. cit.*, I, 32–35; *Columbus*, p. 11.

American coastline first explored by Icelandic colonists from Greenland.[111] Altogether, it would have been difficult to choose a more logical place than Iceland from which to launch a scientific expedition to explore beyond the earth.

Unfortunately for Kepler's purpose, none of this factual background was given or even suggested in the original text of the lunar geography. He assumed his readers would be cultured people, familiar with the content of any books currently available to him. Consequently, he drew upon his very large and diverse library for information which he wove into the lunar manuscript, but he omitted the documenting detail that had determined his selection of material.[112] Then, proceeding as though his audience actually had shared his own gradual psychological conditioning for an extraordinary voyage, Kepler had his mother-image of Ignorance (Fiolxhilde) talk with her precocious offspring Science (Duracotus) about spirits known to dark and distant realms. Listen again to the unearthly ring of Fiolxhilde's words:

> Duracotus, my son, she said, provision has been made not only for the regions you have visited, but for our land, too. For although we have cold and darkness and other inconveniences, which now at last I am aware of when I learn from you the delights of other regions, we are nonetheless well endowed with natural ability, and there are present among us very wise spirits who, finding the noise of the multitude and the excessive light of other regions irksome, seek the solace of our shadows and communicate with us as friends. None of these spirits are especially worthy of note. One, particu-

. . . . . .

[111] Williamson, *op. cit.*, pp. 13–14.

[112] Kepler's footnotes 2, 7, 8, 9, 28, 41, 43, 58, 62, 73, 134, 148, 207, 214, 216, 219, 220, 221, 222, 223. In spite of this enormous range, his footnote 23 betrays incomplete knowledge of Iceland. He underestimated the period of Norwegian hegemony over Iceland by several hundreds of years. Although he knew that the Norse occupation of Iceland had been followed by colonization of Greenland, he was apparently unaware of the voyages southward from Greenland beyond Labrador, or the establishment of eleventh-century bishoprics of the Roman Catholic Church in Markland (Newfoundland) and Vineland (Nova Scotia or New England).

larly friendly to me, most gentle and purest of all, is called forth by
twenty-one characters. With his help I am transported in a moment
of time to any foreign shore I choose, or, if the distance is too great
for me, I learn as much by asking him as I would by going there
myself. Most of what you have seen, or learned from conversations,
or drawn from books, he has already reported to me, just as you
have. I should like you to go with me now to a region he has talked
to me about many times, for what he has told me is indeed
marvelous. . . .

Now it must be kept continually in mind that Kepler's major
objective was to spread word of Copernican science in a way
that would not arouse enemies of science within the church.
By introducing spirits into the moon geography he hoped to
muffle the metaphysicists in their own jargon while at the same
time preparing less prejudiced minds for a wild leap into the
unknown. He took it for granted that scholars of the latter
category would associate the "nine . . . spirits" of Fiolxhilde
with the nine classical muses. When this association failed to
convey itself, he established it in the footnotes[113] by listing the
muses and their counterparts in the primitive science of that
time: "metaphysics, natural science, ethics, astronomy, astrology, optics, music, geometry and arithmetic."

The identity of the spirit most highly regarded by Fiolxhilde,
the one she called the "most gentle and purest," also seemed
self-evident to Kepler in the context of the lunar geography.
After all, the author of the geography was plainly an astronomer, and Fiolxhilde summoned her favorite by calling "twenty-
one characters" into the night. In those days when the Copernican theory was the most discussed subject in astronomy, anyone who could read Latin could surely count the twenty-one
letters in *Astronomia Copernicana!* But these associations also
failed of realization; so ultimately the footnotes spelled them
out, too.[114]

Given the hindsight provided by the footnotes, modern readers cannot help sympathizing with Kepler. How hard he tried to

. . . . . .
[113] Kepler's footnote 35.
[114] Kepler's footnotes 36, 37, 38.

plant advice which his contemporaries could not mistake! Consider what Duracotus said after hearing Fiolxhilde's wish to deliver her precious legacy:

> Straightaway I agreed that she should summon her teacher. It was now spring; the moon, becoming crescent, began to shine as soon as the sun dropped below the horizon, and it was joined by the planet Saturn, in the sign of Taurus, just after sunset. My mother withdrew from me to a nearby crossroads, and after crying aloud a few words in which she set forth her desire, and then performing some ceremonies, she returned, right hand outstretched, palm upward, and sat down beside me. Scarcely had we got our heads covered with our robes (as was the agreement) when there arose a hollow, indistinct voice, speaking in Icelandic. . . .

The stage setting of that passage was astronomically precise, as well as astrologically provocative. The crescent moon does come into conjunction with the planet Saturn in the constellation of the Bull just after sunset. Tycho Brahe had recorded such a meeting in March, 1589. And the last Barents expedition seeking the northeast passage to China had seen another in January, 1600, while wintering in the Arctic.

The covering of the head with robes was a custom Kepler had adopted to shut out light during his observation of an eclipse in the solarium of the Emperor's garden on October 12, 1605.[115] European diplomats had witnessed that unusual public performance.[116] Furthermore, Kepler had dramatized the event by mailing to astronomers in Germany, Italy, Spain, France, Switzerland, and the Netherlands a questionnaire, the answers to which would enable him to compare other observations of the eclipses with his own and so draw broad scientific conclusions.[117] As promoter of this crude forerunner of the modern International Geophysical Year research, he might be forgiven

. . . . . .

[115] Kepler's footnote 49.

[116] Caspar, *op. cit.*, pp. 167–168.

[117] There was particular value in comparative study because the moon, during the 1605 eclipse, was at perigee; Kepler could match the observations against those he had made during the eclipse of 1601, when the moon was at apogee. Caspar, *op. cit.*, p. 167.

for supposing that everybody in science had some familiarity with the episode.

To have such seemingly easy clues to the real meaning of the lunar allegory pass undetected was discouraging enough for Kepler. Infinitely worse was the misunderstanding of what came next in the original text of the geography of the moon. This was the appearance of the "Daemon from Levania." [118]

Kepler's footnotes disclose the extreme care he took to identify the "Daemon." [119] The name itself he took from the Greek. By dropping it unheralded into the midst of the Latin text, Kepler intended to stop his readers at that point and cause them to reflect.

According to modern philology, Kepler was wrong about this, but he believed that the noun *daimon* was derived from the Greek *daiein*, meaning "to know," and that *daêmon* therefore meant "one who knows." He capitalized the word to turn it into a pun, with the alternate meaning of "evil spirit." The enthusiastic acceptance of the alternate is evident from the persecution of Kepler's mother. If his readers had read "Daemon" in the benevolent instead of the malevolent sense (as Kepler, perhaps naïvely, apparently expected them to do) they would have understood that what they were about to read had been assembled by the spirit of knowledge. Modern science would use a different term: specialist.

Without full identification of the Daemon, the real meaning of the allegory might be misconstrued. Kepler therefore did not rely on a Greek word alone to prepare his readers; he coupled the Greek intrusion upon his Latin with another foreign word, this one from the Hebrew: *Levana*, meaning "moon." Ergo, the spirit was a specialist on the moon.[120]

. . . . . .

[118] The Daemon never actually appeared; only its voice was heard. Concerning this, Kepler makes a startlingly modern statement in footnote 50: "I do not consider it impossible, by means of various instruments, to produce individual vowels and consonants in imitation of the human voice."

[119] Kepler's footnotes 34, 51, 72.

[120] Kepler's footnote 42 points out that the more normal designa-

The text of the allegory had already marked this spirit as Fiolxhilde's favorite, and had described how she summoned it by shouting "twenty-one characters" (the letters which spell *Copernican Astronomy* in Latin) into the night. Kepler tells us in the footnotes that he thought his readers would, after deliberation, infer his purpose in writing the manuscript: "to work out, through the example of the moon, an argument for the motion of the earth." [121]

Was it reasonable for Kepler to expect even very intelligent and devoted readers to puzzle their way through his intricate trappings with no more than biographical and etymological clues as guides? The question must remain open because, as we have seen, readers who were not very intelligent penetrated the allegorical masks of Johannes Kepler and his mother Katharina without going further. Lack of an answer is no embarrassment, however, for our purpose here is to understand the content that Kepler intended to convey, whether or not the conveyance succeeded.

Knowing that the voice of the Daemon from Levania is really the voice of a scientist who, through his studies, had become a specialist on the moon as a body governed by the laws of Copernican astronomy, what do we learn as we read the Daemon's contribution to the original text of the lunar geography, amplified by Kepler's footnotes?

We learn first of all that Kepler considered it theoretically possible for men to reach the moon if they were ingenious enough to solve certain problems. His belief in that possibility was implicit throughout the pages that followed. It was a belief that was to become a conviction by the time he told Galileo of the *Dream's* existence in *Conversation with the Star Messenger* in 1610. But already in the *Dream* text itself the problems of a moon voyage were stated succinctly, along with suggested solutions amenable to the laws of physics.

Often there were gaps in his elucidation of his reasoning.

. . . . . .

tion would have been Selenetis; but, he said, the Hebrew word, being less well known to scholars, suited his purpose better.

[121] Kepler's footnotes 3, 96, 125, 131, 135, 146.

From the dimensions of these it is possible to gauge the level of the scientific intelligence to which his original text was addressed. For example, he did not describe the mathematical process through which he arrived at the "fifty thousand German miles" distance from earth to the moon. He simply stated the figure, apparently assuming that the sophisticated readers for whom he had written the allegory would share his own knowledge that calculations derived from the angles of the moon's parallax (its apparent change in position vis-à-vis the earth owing to actual changes in position of its observers on earth) had repeatedly placed the moon 60 earth radii away from the center of the earth, or 59 radii away from the earth's surface.

After learning, from his mother's trial for witchcraft, that he would have to make himself understandable to many who were not prepared to cope with any such abstractions, Kepler put the number of earth radii between the earth and the moon into footnote 53. In the same footnote he gave the length of the earth's radius. Even then, however, he did not break down his message into fragments that could be grasped quickly and easily by a casual reader. There were 15 German miles in each degree of the earth's surface, he said in that footnote; hence the radius of the earth was 860 miles. He did not explain that the 15 German miles per degree had to be multiplied by the 360 degrees in a circle in order to obtain the earth's circumference, that the circumference had to be divided by pi (3.12159) in order to obtain the diameter, and that the diameter then had to be halved to obtain the radius.

How close did Kepler's "fifty thousand German miles" (corrected to 50,740 miles in footnote 53) come to the distance stated in modern statute miles—221,592 at lunar perigee to 252,948 at lunar apogee?

That depends on the correspondence between the modern statute mile and the German mile of Kepler's day. He offers a clue in footnote 53 by comparing the angles of the polestar above the horizon at Rome (40° 50′) and Nuremberg (40° 26′). These two cities, he notes, are virtually on the same meridian. So, he calculates, the overland distance from Nuremberg to Rome is 114 German miles. Jet flight maps of today show the airline distance from Rome to Nuremberg to be 524 statute miles, exactly 4.6 times the German-mile distance stated by Kepler. The old Austrian postal mile was equivalent to 4.6 statute miles. By this measure, the lunar-geography estimate of the distance to the moon totals 233,404 modern miles. But if we accept this figure as the one Kepler intended, then we must attribute to him the belief that a degree encompassed 69 miles instead of the 60 miles we now know to be the correct figure. If, on the other hand, we credit Kepler with an accurate understanding of the mileage in a degree (which would be 4 miles for each of his German miles), the lunar-geography estimate of the distance to the moon would be 202,960 modern statute miles.

Either way, a voyage to the moon would be extraordinarily long, and, as Kepler's allegorical Daemon warned, "the road . . . is seldom open." Nevertheless, the Daemon said, passage could be made by those who were willing to run "grave risk to life" and to season themselves for hardship. Specifically:

"No inactive persons are accepted into our company; no fat ones; no pleasure-loving ones; we choose only those who have spent their lives on horseback, or have shipped often to the Indies and are accustomed to subsisting on hardtack, garlic, dried fish and unpalatable fare."

As an early seventeenth-century forecast of the rigorous training that would be required of cosmonauts in the twentieth century, these opening remarks by Kepler's moon specialist are impressive. The perception behind them is revealed more fully in the statements that "no Germans are suitable" (explained in the footnotes as a stricture against Kepler's gluttonous countrymen)[122] and "we do not despise the lean, hard bodies of the Spaniards" (identified in the footnotes as a compliment both to Spanish frugality and to the Spanish genius for exploring on the basis of astronomical principles[123]—as exemplified by Columbus' westward voyage to reach the Orient).

. . . . . .

[122] Kepler's footnote 61.
[123] Loc. cit.

The fateful words that led to the witchcraft trial of Kepler's mother—"Especially suited are dried-up old crones who since childhood have ridden over great stretches of the earth at night in tattered cloaks on goats or pitchforks"—appeared in the original manuscript at this point, but the footnotes say even they were intended as jocular confirmation of Kepler's belief that, as witchcraft was considered possible, "even this will perhaps be possible, that a body separated from the earth might be carried to the moon." [124]

Having firmly established personnel priorities for the moon voyage, Kepler's lunar specialist next turned to examination of the physical obstacles lying between the earth and the moon. First of these was the radiation showered into interplanetary space by the sun. The footnotes show us the following steps in Kepler's approach to this problem:

1) After being filtered by earth's atmosphere, "sometimes in the summer there is such intense heat from the rays of the sun that forests and wooden buildings catch fire." [125]

2) Experiments with "a glass ball full of water" had demonstrated that "the water transmits the rays of the sun and concentrates them to such a degree that they burn clothing and ignite powder." [126]

3) Even after the sun's rays are reflected from the moon through earth's atmosphere onto the earth, there is warmth in the seeming cool of the silvery glow, which "we can put . . . to proof by the sense of touch." [127]

Kepler must have made this last observation before 1609 because the conclusion is implicit in his original text, which he never recast; in the footnotes he tells of a later experience:

If you catch the rays of the full moon with a concave parabolic mirror, or even with a spherical one, you will feel at the point of focus, where the rays come together, a certain warm exhalation as it were. This happened to me at Linz, when I was concentrating on

. . . . .

[124] Kepler's footnote 60.
[125] Kepler's footnote 200.
[126] Kepler's footnote 223.
[127] Kepler's footnote 200.

other experiments with mirrors and not thinking about the heat of light. I began to look around to see if anyone was breathing on my hand.[128]

4) The atmospheric filter terminates "at the peaks of the highest mountains or even lower down." [129] Above that height, the effects of the sun would multiply ferociously.

From the above progression it was logical to postulate the rapid disintegration of man should he be subject to solar bombardment outside the atmospheric blanket protecting earth. And the Daemon in the lunar geography implied the postulation in prescribing behavior for lunar voyagers after they reach the moon:

"There we betake ourselves hurriedly to caves and to shadowy regions, in order that the sun, coming out into the open a little later, may not overwhelm us. . . . If anything is caught in the daylight, it becomes hard and burnt on top. . . . Whatever clings to the surface is boiled by the sun at midday."

The moon voyagers would have to be protected from the sun, then, throughout the trip. Kepler proposed to solve the shielding problem by using the shadow thrown by the earth as it passed around the sun. If the departure time were fixed to coincide with the moment when the sun stood behind the earth directly opposite the port of embarkation, there would be a tapering cone of darkness like a tunnel sweeping the sky.[130] In this tunnel, Kepler reasoned, the travelers would be safe. His thinking has been confirmed by modern space-probe observations: beyond the range of the scattering effect of earth's atmosphere, more than 99 per cent of the sun's radiation is cut off by the earth's shadow.

Now Kepler understood that because the moon and the earth both were in motion, the shortest route to the moon would not be a straight line between the two, but instead a line from the earth to a point in space at which the moon and the voyagers

. . . . . .

[128] Kepler's footnote 200.
[129] Kepler's footnote 57.
[130] Kepler's footnote 62.

would arrive simultaneously.[131] Either of two courses could be adopted to take advantage of this circumstance.

One was that the voyagers "must . . . circulate aloft for several days in the cone of the earth's shadow" in order to be "on hand at the moment of the moon's entry into this cone." [132] However, Kepler realized that such an arrangement would be "quite contrary and opposed to the nature of the body." There were not only obstacles of physical conveyance to consider but also a question of emotional tolerance; or, as Kepler put it in a footnote: "the disposition of those being transported." [133]

After pondering the problem, Kepler therefore chose the second of his two alternatives: "The whole trip from earth to the moon will be made in that very short time during which the moon is in the cone of the shadow [of the earth]" [134] during a lunar eclipse—and the longest duration of an eclipse is four and a half hours.[135] The lunar specialist in Kepler's allegory stated the solution in these words: "The whole of the journey [between the earth and the moon] . . . is accomplished in the space of four hours at most. . . . We agree not to leave before the eastern edge of the moon begins to go into eclipse. If the moon should shine forth full [that is, if the voyagers should be exposed to the light of the sun] while we were still en route, our departure would be in vain."

"On such a headlong dash," the allegorical Daemon went on, "we can take few human companions—only those who are most respectful of us." [136]

"We congregate in force and seize a man of this sort," the Daemon related; "[and] all together lifting him from beneath, we carry him aloft."

. . . . . .

[131] Loc. cit.

[132] Loc. cit.

[133] Kepler's footnote 63.

[134] Kepler's footnote 62.

[135] Loc. cit.

[136] Kepler's footnote 64 complains that "scarcely one or two . . . philosophers" are willing to take the opportunity that lunar eclipses provide "to extend the boundaries of astronomy" from the safety of earth.

Nowadays we think of flying as a twentieth-century concept. Actually, crude experiments in human flight began a thousand years ago.[137] There was much talk of flying during Kepler's lifetime, and one of the centers of it was at his alma mater, Tübingen University.[138] From his *Conversation with the Star Messenger* we know that Kepler read Della Porta's *Natural Magick*,[139] which described possible methods by which men might propel themselves on artificial birdlike wings. But Kepler knew very well that men would never flap their way through and beyond earth's atmosphere.

In the *Conversation with the Star Messenger* of 1610 he was to speak of a future time when there would be "sails or ships fit to survive the breezes of heaven." In the lunar geography of 1609, however, even mention of sails or ships was suppressed. Although he tells how the propulsive forces will operate, in the allegory Kepler never says what form the propulsion will take. Everything is stated in terms of action by the Daemon, or specialist in moon knowledge. All is theory—pure science. The technology is left to be developed by the spirit of learning.

Kepler's footnote regarding takeoff reveals that the imperial mathematician had constructed a mental model of gravitation and had reasoned out its physical effects.

"I define gravity as a power similar to magnetic power—a mutual attraction," the note says.[140] "The attractive power is greater in the case of two bodies that are near to each other than it is in the case of bodies that are far apart. Therefore, bodies more strongly resist separation one from the other when they are still close together."

The weight of gravity also grows with the mass of the body moving against the gravitational restraint, a companion footnote adds.[141] A ball of lead is harder to move than a ball of

· · · · ·

[137] Lynn White, Jr., *Eilmer of Malmesbury, An Eleventh Century Aviator* (Technology and Culture, Vol. II, no. 2), p. 100.

[138] White, *op. cit.*, p. 104.

[139] *Opera Omnia*, II, 291.

[140] Kepler's footnote 66.

[141] Kepler's footnote 67.

stone "because there is more weight in the former, hence more resistance." Since the bodies of the moon voyagers "are heavy, they will resist motion," and "a quick . . . violent . . . thrust" is prescribed in a further footnote as the only means of breaking free of gravity's grip.

"He [the voyager] is twisted and turned just as if, shot from a cannon, he were sailing over mountains and seas," the allegorical Daemon reports. "The first getting into motion is very hard on him. . . . Therefore . . . he must be arranged, limb by limb, so that the shock will be distributed over the individual members, lest the upper part of his body be carried away from the fundament, or his head be torn from his shoulders."

How this careful arrangement was going to be maintained throughout the voyage was not clear to Kepler. As a theorist, he was unconcerned with technology and entered into no speculations that might have led him to such things as the form-fitting couch occupied by Yuri Gagarin and successor cosmonauts of the twentieth century. In a footnote,[142] Kepler says only that "someone else" would have to "care for the safety of the traveler" in this regard; to ease the rigors, however, the allegorical Daemon did prescribe "narcotics and opiates."

Having already intimated that by the time a real voyage to the moon was made the trip would probably encompass several days, and having explained that the timetable for his dream trip was arbitrarily reduced to four hours because the spirit of knowledge could not at the moment advise another way of keeping the travelers from being shriveled by the sun, Kepler used his four-hour schedule and his estimated distance to the moon (fifty thousand German miles) to further dramatize the rigors of takeoff. In a footnote, he urged his readers to imagine the jolt that would be felt by someone who left the ground at a speed of twelve thousand miles (German miles, that is) per hour.[143]

In yet another footnote, Kepler makes clear his understanding of gravitation as a universal force, operating beyond the earth

. . . . . .

[142] Kepler's footnote 68.
[143] Kepler's footnote 57.

as well as on it.[144] He reiterates an earlier concurrence in much older theories that the pull of lunar gravity is responsible for the rise of tides in earth's surface waters. If the moon thus affects waters in the oceans of the earth, a similiar influence would certainly operate on bodies suspended between the earth and the moon. It follows that if propulsive forces "are imagined to snatch a body aloft toward the vertex of the cone [of earth's shadow] . . . they . . . alone, unaided . . . will labor and sweat and, of course, grow faint. If, however, they undertake the task when the moon is right, the moon, by its presence in the shadow, will aid them . . . by the magnetic attraction of her . . . body."

Hurried reading of those words may leave the impression that Kepler meant to attribute to lunar eclipses a strengthening of the gravitational attraction of the moon. When the passage is placed in the context of the whole work, however, it is plain that Kepler was talking about the forces that would operate on bodies suspended in a direct line between the earth and the moon—any direct line, not merely the tunnel of shadow cast by the earth in passing around the sun. And this is an accurate statement of modern orbital theory insofar as it makes clear the necessity of calculating the opposing pulls of gravity of earth and moon. Furthermore, the "labor and sweat" process is a vivid image of what happens to interplanetary vessels as they leave earth and struggle against the restraint of earth's gravity. The ships go slower and slower as the distance from earth grows until finally, as escape is accomplished, they "grow faint."

En route to the moon, the Daemon noted, the voyagers encountered "a new difficulty; terrific cold and difficulty in breathing."

In dealing with the cold, Kepler explicitly defined the nature of the manuscript he was writing. The cold "we counter with our innate power," the Daemon declared. That this power could be only the human imagination becomes plain in a footnote which adds: the lunar specialist's "burning passion for speculation quite feebly supplies the necessities of life." [145] To

. . . . . .

[144] Kepler's footnote 62.
[145] Kepler's footnote 72.

paraphrase less cryptically: When the demand would become sufficiently great, knowledge would develop technological means of surmounting the scientific problems here being set forth.

The Daemon explained that the moon voyagers' breathing difficulties were offset "by means of moistened sponges applied to the nostrils." Was this perhaps a prediction of the oxygen mask as an artifact of high-altitude travel? Hardly. Joseph Priestley would not identify oxygen until a century and a half had passed. Kepler was simply picking a detail from Aristotle[146] to illustrate a scientific observation: that to reach the moon man would have to carry with him a means of breathing the vital element of life.

Again he had extrapolated from established fact. This time his thought-experiment concerned the density of earth's atmosphere. Like other speculations before and since, Kepler's theoretical construct fell short of perfection. He supposed that the air did not extend very far past the tops of the highest mountains, if indeed it went that far.[147] This, however, was a minor miscalculation alongside his recognition that the atmosphere thins with altitude, that finally the medium becomes too tenuous to retain any heat from sunlight passing through it,[148] and that this thinness is "as fatal to men as lack of water is to fish."[149]

Once far above the atmosphere, the Daemon disclosed a new phenomenon:

"When the first part of the trip is accomplished, the carrying [of the moon voyagers] becomes easier. Then we entrust the bodies to the empty air and withdraw our hands."

If we recognize the Daemon as the spirit of knowledge and accept the analogy that the hands of knowledge are technical instruments grafted upon scientific theory, we can read Kepler's meaning plainly: The moon travelers would no longer need to be propelled; they would be able to float freely.

As this passage appears in the original *Dream* text, it reads

. . . . . .
[146] Kepler's footnote 73.
[147] Kepler's footnote 57.
[148] Kepler's footnote 70.
[149] Kepler's footnote 57.

as though Kepler meant to say that the voyagers thereafter traveled on momentum. But the footnote he appended at this point later suggests a different perspective. Propulsion did cease; but the motion of the travelers ceased, too. The expedition had arrived at a juncture in space where the backward pull of earth's gravity exactly balanced the forward pull of the moon's gravitational attraction, so that the voyagers were "just as if being drawn in no direction at all." [150]

Modern astronomers agree that such a limbo awaits objects that drift into it slowly enough to come to a stop and be trapped. Mid-twentieth century rocket literature contains at least one proposal for an experiment to locate pockets of dense interplanetary dust suspended in this fashion between earth and the moon.

Kepler's moon travelers remained in the free-floating state only long enough to take unto themselves what was left of the energy that had lifted them from the earth. During that momentary pause, they gathered their outstretched limbs together "like spiders" and assumed the shapes of balls. Then the spirit of learning nudged them into reach of lunar gravity "almost by means of our will alone."

Thus Kepler indicated that very little force would be needed to move an object in the rarified "aether" beyond earth's atmosphere. He was anticipating Sir Isaac Newton's third law of motion. In a footnote, however, he emphasized that willpower alone would not do; "there is need of some force also." [151]

Historically, this is a most interesting passage. It dramatizes Kepler's departure from the ancient Greek belief that heavenly bodies were moved by a kind of cosmic intelligence.[152] The will of this undefined authority was enough to account for any aberration of stars, planets, or comets which could not readily be rationalized otherwise. By substituting a physical force for

· · · · · ·

[150] Kepler's footnote 75.

[151] Kepler's footnote 76.

[152] In his *Cosmic Mystery* of 1597, Kepler still had referred to the "soul" of a planet. See Max Jammer, *Concepts of Force* (Cambridge, Mass.: Harvard University Press, 1957), pp. 81, 82.

the metaphysical will,[153] Kepler not only denied the Greek tradition of cosmic intelligence but called attention to the existence of a new condition he was first to refer to as inertia.

Aristotle had taught that moving objects were propelled by the air about them, and that movement ceased with cessation of the motive power. In Paris, during the fourteenth century, a scientific school headed by Jean Buridan drew upon practical experience to show that, instead of propelling, the air actually resisted an object passing through it, and that the resistance seemed to vary with the size of the object and the impetus behind it.[154]

For some years before writing the lunar geography in 1609, Kepler had been struggling toward a clearer understanding of motion than Buridan's theory provided. According to Max Jammer, an Israeli scholar who explored this question more deeply than anyone since Max Caspar, the turning point came in 1606,[155] the year after Kepler's discovery of the elliptical form of planetary orbits.[156] In demolishing the ancient notion of circular orbits, Kepler had to find some way of explaining why planets would travel elliptical paths, for Plato, believing that circular motion was perfect, had declared dogmatically that the heavenly bodies could move only in circular paths.

· · · · · ·

[153] Kepler explicitly confirmed the substitution in the second edition of *Cosmic Mystery*, published in 1621, the year of the beginning of the lunar geography footnotes. (Jammer, *Concepts of Force*, p. 90.)

[154] Thomas S. Kuhn, *The Copernican Revolution* (Cambridge, Mass.: Harvard University Press, 1957), pp. 118–121.

[155] Six years before Galileo suggested the inertial principle in restricted form in his *Second Letter on the Sunspots*. See Erwin Panofsky's mistaken conclusion on this point in *Galileo as a Critic of the Arts; Aesthetic Attitude and Scientific Thought*, Isis, XLVII (1956), cited by Hellman in her English translation of Caspar's *Kepler*, pp. 136–137

[156] Though not published until 1609, the discovery was first announced in a letter from Kepler to David Fabricius. See *Gesammelte Werke*, XV, 240–280, cited by Hellman in her translation of Caspar, *op. cit.*, p. 170.

Without yet using the word "inertia," Kepler said, in discussing the nova of 1604,[157] that the weight of the planets appeared to account for their resistance against following circular courses. In publishing the elliptical discovery of 1609 in *New Astronomy*, he went further, saying that the mass of a body would determine its dynamic response to the gravitational attraction of another body.[158] In the lunar geography, the concept of inertial mass was taken for granted and dramatized in the behavior of the bodies of the moon voyagers.

"Every body by reason of its own matter has a certain inertia in regard to motion which provides repose to the body whenever the body is placed in a position beyond powers of attraction," a footnote to the geography explains. "Whoever would move this body from its place must overcome this force, or, rather, this inertia." [159]

Here the modern meaning of the word "inertia" is clear. But the footnote was not written until 1621 or later. By that time, Kepler already had used the word repeatedly in his *Epitome of Copernican Astronomy* and had stated the proposition that, were it not for inertia, the planets would acquire infinite velocity from the smallest motive force.[160] This, of course, was a statement only of the inertia of rest. Not until Sir Isaac Newton added the inertia of motion would science at last have the concept that, in the absence of intervening force, a resting body remains at rest and a moving body continues in motion.[161] Nevertheless, it was Kepler who made the beginning,[162]

. . . . . .

[157] *Opera Omnia*, Vol. II (*On the New Star in the Constellation of the Serpent*), quoted by Max Jammer, *Concepts of Mass*, p. 54.

[158] Jammer, *Concepts of Mass*, p. 54.

[159] Kepler's footnote 76.

[160] Jammer, *Concepts of Mass*, p. 55.

[161] Sir Isaac Newton, *Mathematical Principles of Natural Philosophy*, translated by Andrew Motte and revised by Florian Cajori for the University of California Press, 1934 (Chicago-London-Toronto: Great Books of the Western World, 1952), vol. 34, p. 14. Quoted in I. Bernard Cohen, *Birth of a New Physics* (New York: Doubleday, Science Study Series S 10, 1960), p. 157.

[162] For a full discussion of this point, see Jammer, *Concepts of Mass*, pp. 49–64.

and one of the forgotten steps in that beginning was the lunar geography of 1609.

After Kepler's moon expedition entered the outermost fringes of lunar gravity, the Daemon reported that "finally the corporeal mass heads of its own accord for the appointed place." The pace was too slow to complete the trip within the protection of the shadow cone of the eclipse, so "we hurry the bodies along with willpower."

As the travelers approached the moon at rapidly accelerating speed under the growing influence of the moon's gravitational attraction, the Daemon had to "precede [the voyagers] . . . lest some damage be inflicted by a very hard impact on the moon." Did Kepler thus imply a braking rocket? No more than he provided a rocketship for the takeoff. Once again leaving the development of technology to the spirit of learning, Kepler merely dramatized the necessity of slowing the moon-landing if man was to arrive unhurt. At the end of the voyage he sidestepped description of the landing by having the Daemon say, "when we disembark as it were."

Having drugged his voyagers immediately before their departure from earth, Kepler had to waken them upon arrival at the moon. "They usually complain about the unspeakable weariness of all their limbs," the Daemon reported, but "they later entirely recover from this, so that they can walk." [163]

Seen in hindsight from the twentieth century, Kepler's theoretical guidance of an early seventeenth-century lunar expedition across the physical obstructions that modern astronomers recognize as lying between the earth and the moon was an incredibly prescient exercise. Rarely have scientific problems of future generations been so faithfully set forth. But to Kepler

. . . . . .

[163] Kepler's footnote 79 emphasizes his belief that men could travel to the moon if they made the proper preparations. "Let the traveler see to it that he arrives in such an unharmed condition that he can awaken," the footnote says. Footnote 80 then points out that the whole business of traveling through a tunnel of shadows in the sky is not natural to the human animal, implying that some vehicle would have to be found in which men could ride through interplanetary space without fear of being destroyed by the sun.

all this was merely means to a larger end. His major purpose in writing the lunar geography was to demonstrate to his fellows that the human animal was not the central figure of the cosmos, that the heavens did not wait upon his home planet, earth.

Kepler's lifelong hero, Copernicus, had fallen considerably short of a convincing argument in composing his thesis that the earth moved around the sun. The heliocentric idea was not widely accepted in 1609, even among scholars. If the earth moved (so popular thinking ran), people surely would be thrown off into the air or thrown down upon the ground. However, anybody with eyes to see the sky could see that the moon moved. Therefore, Kepler reasoned, by taking people vicariously to the moon and standing them seemingly still there, he could show them the earth in motion.

He set the stage for this primary task as soon as his moon travelers were tucked away in places safe from the sun. His first piece of preparation was psychological: a new name for the earth. Volva, he called it, to designate the revolving planet.[164]

If there remained any doubt that Kepler had written the original text of the lunar geography exclusively for scientists, the doubt would be gone at this point in the manuscript. For he assumed his readers to be sufficiently familiar with astronomy to know that one hemisphere of the moon always turned toward earth.[165] He named this hemisphere Subvolva in recognition of its continuous presence under Volva's gaze, and called the other hemisphere Privolva to designate its eternal deprivation of the glorious sight of Volva.

Next he offered a detailed analysis of the apparent motion of the sun and other stars around the moon, comparing these to the apparent motion of the sun and stars around earth. In the footnotes, he explains that he expected it to be understood that the motions so described were really the motions of the moon and the earth, unconsciously transferred by human observers deluded by the seeming immobility of their footing.[166]

. . . . . .

[164] Kepler's footnotes 89, 90.
[165] Kepler's footnotes 88, 92, 127.
[166] Kepler's footnotes 110, 180.

The geography text mentioned in passing that the duration of lunar days and nights fluctuated slightly. In the footnotes, we find that Kepler anticipated understanding by his readers that these variations were due to the orbital ellipses he had discovered—that at times the passage through space would be slower than at other times.[167] He likewise expected his audience to grasp that, when the geography spoke of Mars and Venus occasionally looking twice as big in the lunar sky as they did above the heads of earthly observers, he again was referring to relative positions brought about by elliptical orbiting.[168]

So engrossed, in fact, did Kepler become in the scientific aspects of his disguised lunar geography that he neglected to be consistent in the fictional aspects of his *Dream*. His Daemon, for example, was introduced as a lunar being. But once the description of affairs on the moon was launched, the Daemon often talked as a creature native to earth rather than as a visitor to it, repeatedly referring to terrestrial phenomena as "our" and to lunar phenomena as "their." Although these unconscious crossovers from the third to the first person jar the eye, prepared readers will not find them difficult to follow. Now and again Kepler's dedication to scientific exactitude does bog the flow of his text momentarily. Generally, however, the moon geography is lit by sprightly imagination.

For example, the names Subvolva and Privolva need only be pronounced to convey the meaning behind them. But Kepler's purpose in composing the geography required that the motion of Volva be sensed. He provided the visual illusions by identifying Volva's geographical features, as these were discerned from Subvolva, in simple analogies to what earthlings see of lunar geography and know as "the face of the man in the moon." South America, for example, looked like a bell swinging from a rope.[169] Europe appeared as a girl silhouetted in a flowing dress, about to kiss a lover, her arms outstretched behind her.[170]

. . . . . .

[167] Kepler's footnotes 111, 114.
[168] Kepler's footnote 121.
[169] Kepler's footnotes 167, 168.
[170] Kepler's footnotes 159, 160, 161.

These images appeared to Subvolva in the same regular procession throughout the night. They served as substitutes for the hands of a clock in measuring intervals of time. One complete procession occupied twenty-four earthly hours. There were fifteen processions in one Subvolvan night.

Travelers in the Subvolvan hemisphere could also tell where they were on the moon by looking at Volva. Volva always hung over a given place at a fixed angle, as though nailed to the sky at that point: directly overhead here, close to the horizon there, to the right of this mountain, in the cleft of that valley. These effects were, of course, due to the unchanging position of one half of the moon in relation to earth.

The sunlight reflected from Volva was a glorious sight to Subvolva because of Volva's immense size—four times the breadth of the moon, fifteen times its reflecting surface. To those who saw it on the lunar horizon, Volva looked like a mountain on fire far away. The light itself was strong enough to be felt appreciably as heat; it moderated the chill of the Subvolvan night.

At intervals, from Subvolva, a small black spot rimmed with red could be seen to cross the face of Volva. It traveled faster than the geographical markings that told the passage of time. This nimbler spot was the shadow of the moon eclipsing Volva.

At other times, the face of the sun would be eclipsed by Volva. On some of these occasions, Volva would momentarily acquire a luminous halo. On other occasions, the sun's light would be utterly extinguished and Subvolva would be plunged into Stygian dark.

The rare total eclipses were the only periods when Subvolva was without some degree of illumination from Volva. Even in the Subvolvan day, when the sun dominated the sky, Volva was visible because of its size, although it waxed and waned as the moon does to watchers on earth.

Never able to see Volva, the Privolvan hemisphere had no light at night except that of the stars and of the planets other than earth. Except for the occasional blazing-up of Venus and Mars, twice as big as earth ever sees them, the Privolvan night was an abysmally dark and gloomy period lasting the length of fifteen days and fifteen nights on earth. This was the duration of time which Kepler, as a terrestrial astronomer, knew the Subvolvan half of the moon to be steadily reflecting the light of the sun toward earth. Since Privolva faced the other way, it had to look into a sky bereft of sunlight throughout those two earthly weeks. The Privolvan weather during that time was bitter cold, with icy winds. During the following two weeks the pitiless glare of a great, slow-moving sun baked the hemisphere in heat fifteen times as intense as that of the Sahara desert.

What Privolva looked like, Kepler had no way of knowing. But he assumed the terrain must be wild because of the extremes of temperature to which it was subjected. The surface of Subvolva he had studied intensively for years; from the shadow studies Professor Maestlin had taught him to make, he was confident it had "very high mountains and very deep and broad valleys, and thus falls far short of earth in perfect roundness. Moreover, the whole of it . . . is porous and pierced through, as it were, with hollows and continuous caves. . . ."

No explanation of the Maestlin shadow-study technique is to be found in the text of the geography. But there is a footnote[171] in which Kepler describes the method, attributes it specifically to Maestlin, says it was part of the 1593 theses, and tells how it was applied during an eclispe of the sun in the year 1612: First, the image of the sun was projected through a telescope onto a sheet of white paper. Then, on the paper, and as the eclipse progressed, it was possible to follow the globe of the moon during its passage across the sun's face. On the advancing edge of the lunar disk, two small bumps were noticeable. These were watched until it could be ascertained that they moved at the same speed as the moon. Thus the bumps were established as promontories on the moon's surface. Their height was estimated by comparing the breadth of their shadow to the breadth of the whole moon's shadow. The resulting altitudes were considerably above those of earthly mountains.[172]

. . . . . .

[171] Kepler's footnote 207.

[172] In his footnote 207, Kepler fixes the diameter of the moon at five hundred miles and the height of the lunar mountains at eight miles.

In the midst of an otherwise remarkably modern description
of the moon's geographical features, Kepler committed what
most (but not all)[173] modern scientists would call a spectacu-
lar mistake: He covered large parts of the lunar surface with
seas comparable to the oceans of earth. He then populated these
waters with creatures in thick, porous, husklike skins of a sort
that astronomers would agree to be necessary to prevent death
from dehydration in the near-vacuum of the lunar atmosphere.[174] And he gave the bodies of these moon creatures aqua-
tic and serpentine forms appropriate to survival in a network of
rivers and lakes running through subsurface caverns.[175] Finally,

. . . . . .

[173] For an exception, see V. A. Firsoff, *Strange World of the
Moon* (New York: Basic Books, 1959), p. 171.

[174] The question of life's origin on earth fascinated Kepler, as it
did many other men of science of his time. The belief that living
creatures could arise spontaneously from nonliving matter (a concept that is only now gaining experimental documentation) was
being heatedly debated about him, and he reflected his involvement
in the argument by postulating the emergence of life forms from
sun-charred debris on the moon in the cool of evening. Footnote 221
supports this theoretical excursion by quoting from the philosopher
Julius Caesar Scaliger. Scaliger had chronicled sailors' tales of ducks
that took form from resin sweated by the sun from the planking
of ships; the bills of these birds formed last, and upon their completion the ducks dropped into the water and floated away. Kepler
also reported his own observation of a twig of a tree from which
an insect seemed to be rising from an excrescence of sap.

[175] Adaptation of species, we see from this passage, seemed as
logical to Kepler as it did to Charles Darwin several hundreds of
years later when Darwin began to put together his theory of evolu-
tion. In footnote 213, Kepler draws the even more modern conclusion that life is governed by fixed cycles and rhythms.
"For us here on earth, the very slow motion of the fixed stars, the
brief orbital periods of the individual planets, and the daily revolution of the earth itself seem to me to be related to the length of
human life and moderate size of the human body," this footnote
says. Therefore, "since for the moon the fixed stars return more
quickly than Saturn, and since the day is thirty times longer than

he used this land of watery caves as a springboard to jump to the loftiest peak of insight of his life.

During the Subvolvan day, Kepler's Daemon pointed out, the sun and Volva both were visible in the sky. Their gravitational attractions

. . . in combination lure all the water to that hemisphere, and the land is submerged, so that very little stands up above the water, while the Privolvan hemisphere, on the other hand, is dry and cold, since, of course, all the water has been drawn away. When, however, night approaches the Subvolvan area, and day comes to Privolva, since the luminaries [sun and Volva] are divided between the hemispheres, so also is the water, and the fields of Subvolva are laid bare, but moisture is provided to Privolva as a slight compensation for the heat.

Now historians generally have attributed formulation of the universal law of gravitation to Newton. The gist of the law, as he phrased it, was the proportion between the mass of a body and the power of the body's gravitational attraction (earth and the falling apple; earth and the hovering moon). Newton's formulation is dated variously from 1666 to 1686.

In Kepler's lunar geography, dated 1609, we find first the pull of the moon drawing the waters of earth upward to create tides, then the contending pulls of the earth and the moon on the moon voyagers, then the combined pull of the earth and the sun on the waters of the moon engulfing a whole lunar hemisphere, and finally the pull of earth alone, counter to the pull of the larger but more distant sun, removing the water from the moon's flooded half.

. . . . . .

ours, I thought that I should ascribe to lunar creatures a short life and a fast rate of growth."

It is worth noting that in several footnotes Kepler raised doubt not only about whether moon creatures would be similar to earth's humans, but also about whether the moon would have on it any life at all. "If you concede that there are living beings on the moon," he said (footnote 115), and "Since they inhabit it, as we are now imagining" (footnote 125), and "Lunar dwellers, if there are any" (footnote 146).

Here certainly is a dramatization of the universal nature of gravity,[176] half a century before Newton.

Sir Isaac acknowledged that his laws of motion were derived from Kepler's planetary orbit laws. How much did Kepler's lunar geography contribute to Newton's law of gravity?

Whatever the ultimate verdict of historians, it can be said now that in writing the first geography of the moon, Johannes Kepler performed a feat of scientific imagination unmatched before or since.

VI

About the correct interpretation of the main body of Kepler's moon geography there seems no reasonable doubt. But the opening passage of the manuscript may have been intended to say much more than has been attributed to it. Here again are the words:

In the year 1608, when quarrels were raging between the brothers, Emperor Rudolph and Archduke Matthias, people were comparing precedents from Bohemian history. Caught up by the general curiosity, I applied my mind to Bohemian legends and chanced upon the story of the heroine Libussa, famous for her magic art. It happened then, on a certain night, that after observing the stars and moon, I stretched out on my bed and fell sound asleep. In my sleep I seemed to be reading a book I had got from the market. . . .

Because of the quarrel between the Hapsburg brothers, the people of the ancient eastern European kingdom of Bohemia were searching the past for precedents. What sort of precedents?

. . . . . .

[176] In his extended analysis of Kepler's contribution to gravitational theory, Max Jammer (*Concepts of Force*, p. 84) interprets a letter Kepler wrote in 1607 to mean "that Kepler had already conceived the universal character of attraction, an idea generally attributed only to Newton." Jammer conceded, however, that the letter referred to could be read in a more limited sense. The lunar geography text would appear to confirm Jammer's broader conclusion, which credits Kepler with awareness that gravitational attraction could be expressed in mathematical terms.

Two and a half centuries before Kepler wrote the geography of the moon, during the reign of the Bohemian King Charles (1346–1378), this little country had become the fountainhead for ecclesiastical reform movements. The Bohemian John Huss was burned at the stake on July 6, 1415, for his part in those agitations. His posthumous revenge was a civil war that ended in compacts guaranteeing the Bohemian people freedom to worship as they as individuals chose.

The Papacy, never content with such an arrangement, saw a chance to revise it to Rome's liking when Rudolph II assumed the throne in 1576. In 1600, the year when Kepler fled to Prague to escape Catholic persecution in Graz and to work with Tycho Brahe, the Pope's envoy was urging the childless Rudolph to take steps to protect the Hapsburg succession. Rudolph's brother, Matthias, used the religious controversy to further his own political ambitions. Finally Matthias marched troops into Bohemia in defiance of Rudolph's authority. On June 25, 1608, Rudolph made temporary peace with his brother by agreeing that Matthias should rule Hungary, Moravia, and Upper and Lower Austria, leaving Bohemia and neighboring Silesia alone of the Hapsburg empire in Rudolph's hands.

Rudolph's imperial mathematician could not have foreseen that within three years the Emperor would be dethroned by Matthias, or that Matthias in turn would engulf Europe in wolf-pack war for thirty years (1618–1648) by trying to transfer rule to his cousin Ferdinand, a Romanist fanatic. Kepler was close enough to the Emperor, however, to know that the squabble with Matthias was not just a minor difference of opinion within the royal family. Kepler was aware that the clashes between the brothers involved politics and property holders all over the Continent. Kepler's biographer, Max Caspar, tells us, without explaining the statement: "Kepler took an active part in these events." [177]

Is it possible that the lunar geography was an element in this activity—that beneath his desire to broaden understanding of the Copernican theory was a hope on Kepler's part that the new

. . . . . .

[177] Caspar, *op. cit.*, p. 204.

understanding might help to stave off the war that was brewing in 1608?

Let us assume that the author of the geography had such an idea. The subject would be too explosive to discuss in any detail outside diplomatic dispatches. The most that Kepler would dare to do would be to throw out a few broad hints and let his readers chew on those. The goings-on between Rudolph and Matthias were familiar to every well-educated man in Europe. Kepler would not need to explain the background to them; he need only chose a provocative suggestion to sharpen their awareness.

Had this indeed been his strategy, it could explain the brief appearance in the lunar geography of the character named Libussa. Libussa took no further part in the book. Kepler merely introduced her name before falling asleep and dreaming of reading about the voyage to the moon.

Who was this Libussa? "Famous for her magic art," Kepler said of her. That's all he said, except to characterize her by the Latin word *virago*. In modern usage, virago has come to mean a termagant, a shrew, a noisy and unpleasant woman. But in Kepler's day a virago was a female of heroic attributes. If this old definition is accepted, it can be seen that Kepler once again was punning when he wrote of Libussa's "magic art." For a manlike woman named Libussa reputedly founded the city of Prague. She was the primal mother-figure of the Bohemian people, older than the Hussites, older than the Hapsburgs.

According to the legends which the lunar geography said Kepler was reading when he fell asleep that night in 1608,[178] Libussa was the daughter of old King Krok, or Crocus. She may or may not have been a real person. If she lived, her rule may have ended in the year A.D. 505, or in A.D. 588, or in A.D. 734. The ancient chroniclers differ about this, as they do about her name, which may have been spelled Libussa, Lobussa, Lubossa, Lubussa, or Lybussa. But whether she really existed in flesh and blood or was only a figment of patriotic imagination, she lived

. . . . . .

[178] Johann Heinrich Zedler, *Grosses Vollstandices Universal Lexicon* (Halle and Leipzig, 1738), vol. 17, p. 809.

in the Bohemian soul and her importance derived from her womanhood.

The legends say that Libussa began her reign as a maiden queen. She governed alone with wisdom and courage. Her romantic people were unhappy with her lonely state, however, and for their sake she sought a husband. But either she mistrusted her judgment in men or else she didn't see enough difference among the eligible prospects to make a choice worth her time, for she turned a white horse loose in the fields after committing herself to wed whichever man the animal stopped in front of. She expressed the hope that the chosen person would be sitting at an iron table (as a queen she had a right to dream of a gentleman of fashion), and the horse stopped in front of a fellow who was eating lunch on a plowshare. True to her bargain, Libussa married the plowman, whose name was Premislaus. With him as a sire, she mothered the long line of Premyslide princes, including good King Wenceslaus of the Christmas carol.

Undeniably a woman, yet endowed with all the positive virtues of a man, this Bohemian queen, in days when other women were counted little more than cattle, would have seemed to depend on magic for her command of people's loyalty. Kepler would have been justified in making one of his typical quips about her extraordinary powers, for her gift of prophecy was such as to bring her the nickname, Sibylla Bohemica. And she would have been the one universally adored folk-heroine whose name he might have invoked with some reasonable expectation of extending the truce that was arranged in 1608 in the quarrel between the Hapsburg brothers.

Contrivance is seldom without some particular meaning. What meaning was there behind Kepler's contrivance of a seemingly meaningless shift from the opening scene of the lunar geography in which he mused about Libussa, to a scene centered on another mother-figure, Fiolxhilde?

We now know, from Kepler's footnotes, that this second mother was dual: Kepler's own mother and the symbolic mother of scientific knowledge. Regarding the latter, Kepler wrote into a letter his conviction that "if there is anything that can bind the heavenly mind of man to this dusty exile of our earthly home and can reconcile us with our fate so that one can enjoy living—then it is verily the enjoyment of . . . the mathematical sciences and astronomy." [179]

No violence is done to Kepler's meaning by interpreting this passage to mean that if men, by observing the movements of the cosmos about them, wholly realized the odds against themselves in the grip of the vast forces shaping their environment, they would stop wasting their energies and talents on petty squabblings and concert their efforts toward understanding the environment—what Kepler would have called "The Glorification of God."

In the long perspective of human history, the position of God is the foremost difference between the seventeenth and twentieth centuries. Where men of Kepler's day depended on God to guarantee intelligent social behavior, the modern tendency is to throw responsibility on man's capacity to govern himself. Once the facts have been adjusted to fit the new emphasis, Kepler's idealistic approach to the practical realities of 1608 no longer seems farfetched. Indeed, it parallels the argument supporting scientific bridges across the religious gulf separating twentieth-century Russia and America. As Lloyd Berkner put it during his chairmanship of the Space Science Board of the U.S. National Academy of Sciences, exploration of the cosmos "provides a new perspective from which to view the more petty differences among men on Earth and to focus in better proportion on the more critical aspects of our existence." [180]

Among the lunar geography footnotes are several references lending credence to this line of speculation. In one of these notes the imperial mathematician discloses that political arguments were included in the 1593 Tübingen theses from which the moon dream arose, and hurriedly he lists some of the sociopolitical meanings he had intended to be read into the original

. . . . . .

[179] Baumgardt, *op. cit.*, p. 190.

[180] Lloyd Berkner, *Geography and Space*, an address to the American Geographical Society in New York City on January 20, 1959.

geography manuscript.[181] Apparently he had set one whole footnote aside for elucidation of those hidden meanings. Unfortunately, all that is left of that particular note is its number.[182]

Two surviving pieces of Kepler's personal correspondence add support to the notion that his lunar allegory originally had a political as well as a scientific purpose. Writing to Matthias Bernegger on December 4, 1623, he asked: "Would it be a great crime to paint the cyclopian morals of this period in livid colors, but for the sake of caution, to depart from the earth with such writing and secede to the moon?"

By the time he set it down, the question had become rhetorical. His mother's trial had killed her and taught him to be wary. In the next sentence he told Bernegger that the idea already had been abandoned. "What good will it do to flee?" the letter continued. "Neither More in *Utopia* or Erasmus in *Praise of Folly* was safe. Therefore, let us dismiss this political muck entirely and remain in the charming groves of philosophy."

Five years and three months later, on March 2, 1629, Kepler wrote to Bernegger again: "What if I submit to you, for fun, my Lunar Astronomy, or heavenly phenomena in the Moon?"

Astronomy. Nothing more. The Thirty Years' War was then eleven years old. If peace came now, it would have to find its own way.

In December of 1629, Kepler began setting the lunar geography in type. By the following April, six pages of the manuscript had been printed. Then his pocketbook was exhausted. Leaving his family in Sagan, he somehow reached Leipzig, where he borrowed fifty florins and bought a half-starved horse. He rode this pitiable animal through a soaking rain to Ratisbon, where the German Reichstag was in session. He hoped to appear before the legislators there and appeal for the right to sell some impounded Austrian bonds. He was hardly off the horse before he was down in bed with a fever. On November 15, 1630, he died.

After three days—Kepler would have been amused by its hap-

. . . . . .

[181] Kepler's footnote 82.
[182] Kepler's footnote 83.

pening one day before lunar eclipse—the body of the imperial mathematician was lowered into a grave in the Lutheran churchyard in front of St. Peter's Gate. Later, the war he had tried to discourage wrecked the cemetery. Even the tombstones were demolished, so that no one now knows the last resting place of his bones.

The geography of the moon might never have been published had Kepler left his widow and children in comfortable circumstances. Only because of their poverty was the manuscript printed and sold four years after Johannes' death. His son-in-law and his son edited the copy with no conception of the scientific genius reflected between the lines. For three and a quarter centuries the real message of the document has lain almost continuously buried in library dust. At no time has this incredibly prescient work of Kepler's remotely approximated the fame of the words his English contemporary, Will Shakespeare, put into the mouth of a character (momentarily posed as "a sectary astronomical" to tell "what should follow those eclipses") to express what I suspect Kepler intended the lunar geography to say:

This is the excellent foppery of the world, that, when we are sick in fortune—often the surfeits of our own behaviour—we make guilty of our disasters the sun, the moon, and stars, as if we were villains on necessity, fools by heavenly compulsion, knaves, thieves, and treachers by spherical predominance, drunkards, liars, and adulterers by an enforc'd obedience of planetary influence, and all that we are evil in, by a divine thrusting on. . . .[183]

. . . . . .

[183] *King Lear*, Act I, Scene II, in *The Complete Plays and Poems of William Shakespeare*, edited by W. A. Neilson and C. J. Hill (Cambridge, Mass.: The Riverside Press [of Houghton Mifflin Co.], 1942).

The Dream

Note: Occasionally the footnote numbers in this translation are not the same as those in the original, partly because of the difference in word order in the two languages, and partly because in the original itself the numbers were out of sequence. It was felt that it would be less confusing to the reader to renumber these few notes than to leave them as they were. The following numbers have therefore been transposed: 3 and 4, 209 and 210, because of translation problems; 170 and 171, because of the original error in numbering; and 33 and 34 in the appendix, because of translation problems.

Frisch's *Opera Omnia*, Volume VIII, was the source document for this work.

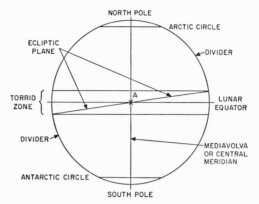

A diagram of the moon, illustrating the terms used by Kepler. A is the point Volva (earth) is directly above at all times. From any point on the divider, Volva is always seen on the horizon. (*Courtesy of M. W. Makemson.*)

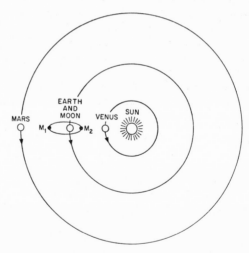

Orbits of Mars, earth, moon, and Venus, showing that the moon is sometimes closer to Mars or Venus than the earth is. (After Kepler.—Diagram is not to scale.) M_1 is the moon at apogee; M_2 is the moon at perigee. As the orientation of the moon's orbit continually varies, in four and a half years apogee will bring the moon still closer to Venus and perigee toward Mars. The orbit of Mercury between Venus and the sun is not shown. (*Courtesy of M. W. Makemson.*)

THE DREAM

by Johannes Kepler,

THE LATE IMPERIAL MATHEMATICIAN

A Posthumous Work on Lunar Astronomy

PUBLISHED BY HIS SON LUDWIG KEPLER, M.A.,
CANDIDATE FOR A DOCTORATE IN MEDICINE

Printed in part at Sagan in Silesia,
completed at Frankfurt,
at the expense of the author's heirs

1634

To *the Most Illustrious and Exalted Prince and Lord,*
Philip, Landgrave of Hesse
Count of Katzenellenbogen, Dietz, Ziegenheim,
and Nidda, etc.

To His Most Clement, etc., Lord and Prince.

Most Illustrious and Exalted Prince, Most Clement Lord:
When my father, Johannes Kepler, the Imperial Mathematician, had become exhausted by the motion of this earth, he
began to dream about the astronomy and the motion of the
moon. I know not what omen this dream brought with it!
It certainly was very distressing to us, his children, even
though the outcome has, for him, been agreeable enough and
even most desirable. For when this *Dream* had just been written and was about to be printed, my father, taken with a
heavier sleep (alas!), indeed a mortal one, flew in spirit to
the heaven above the moon (we hope), and abandoned us,
his children, to be exposed to the wounds of war and the
miseries of this world, almost destitute of earthly wealth. My
most renowned and learned brother-in-law, Jacob Bartsch, doctor of medicine and professor designate of mathematics in
the Strasbourg Academy, undertook the task of printing. But
he also was overtaken by a fatal disease and died, leaving the
work still unfinished.

I, meanwhile, returned to Germany from a trip with a certain Austrian Baron. Having had no news of the condition of my relatives for two years, I wrote to them in Lusatia, from Frankfurt, asking that they let me know whether they were still alive and how they were. Behold, my stepmother, an impoverished widow with four orphans, came to me in a turbulent time, to a place most unsuitable because of the costliness there of the necessities of life. And she brought with her incomplete copies of this *Dream*, and implored aid of me who am myself in need of assistance and advancement from others. She asked me to complete the copies of this *Dream*. But what good can I hope will come of it, since it proved fatal to my father and my brother-in-law? Nevertheless, it befits a son not to hide his father's famous and honored name but rather, if he cannot increase that reputation with his own genius, to preserve it as best he can. I could not refuse the request, and in fact I have acceded to it.

Thus far, a patron for the undertaking is lacking. One will hardly be found among the military, since they are little concerned with the astronomy of the moon but rather must bestir themselves lest they be wounded or crushed by cannonballs or other missiles. Wherefore, I could find no one more worthy than you, Most Illustrious Prince, whose patronage this work might enjoy. You are most proficient in mathematical studies, you are far removed from the tumult of war, and you fostered my father with your very kind patronage while he was still living. Hence the orphans, borne up by their trust that you will not deny your patronage to them and to this work, through me most humbly commend themselves and this *Dream* to your most Illustrious Highness. With most zealous prayers they beseech almighty God to preserve your Most Illustrious Highness, along with your Most Illustrious Spouse, sound in body and mind, and to ward off all enemy attack and military threats from your kingdom.

May you, therefore, Most Exalted Prince, flourish long for God and Country.

Frankfurt on the Main, September 8, 1634.

I am, Most Illustrious Highness,

Your most devoted Servant,

LUDWIG KEPLER, M.A.,

CANDIDATE FOR A DOCTORATE IN MEDICINE.

In the year 1608, when quarrels were raging between the brothers, Emperor Rudolph and Archduke Matthias, people were comparing precedents from Bohemian history. Caught up by the general curiosity, I applied my mind to Bohemian legends and chanced upon the story of the heroine Libussa, famous for her magic art. It happened then on a certain night that after watching the stars and moon, I stretched out on my bed and fell sound asleep. In my sleep I seemed to be reading a book I had got from the market. This was how it went:

My name is Duracotus.[1] My home is Iceland,[2] which the

. . . . . .

[1] The sound of this word came to me from a recollection of names of a similar sound in the history of Scotland, a land that looks out over the Icelandic ocean.

[2] It means "land of ice" in our German language. I saw in this truly remote island a place where I might sleep and dream and thus imitate the philosophers in this kind of writing. For Cicero went over to Africa to dream. And in that same western ocean Plato fashioned Atlantis, whence he summoned mythical support for military valor. Finally, Plutarch, in his book about the face on the moon, after much discussion digresses to the American Ocean and describes an arrangement of islands which some modern geographer will probably identify as the Azores, Greenland, and Labrador, regions situated around Iceland. As a matter of fact, every time I reread that book of Plutarch's I wonder greatly how

. . . . . .

it happened that our dreams, or rather our fictions, were in such close agreement. I have a perfectly trustworthy memory, and I recall the circumstances of the individual parts of my story, inasmuch as not all of these followed from a reading of Plutarch's book. I still have an old paper written by your hand, most illustrious D. Christoph Besold, when in 1593 you had taken from my dissertations about twenty theses concerning the heavenly phenomena of the moon and, prepared to argue them if he had given assent, you had shown them to D. Vitus Muller, who was then the regular chairman of philosophical debates. At that time I had not yet seen Plutarch's works. Later I chanced upon the two books of Lucian's *True Story*, written in Greek. I selected those books as a means of learning the language. I was aided by the enjoyment of his very lively tale, which nevertheless gave some hint of the nature of the whole universe, as Lucian himself points out in the preface. Lucian, too, makes a voyage beyond the Pillars of Hercules into the ocean and is snatched aloft, ship and all, by a whirlwind that carries him to the moon. These were my first steps in the trip to the moon which I pursued at a later time. In 1595, at Graz, I first got hold of Plutarch's book, which was brought to my mind by my reading of Erasmus Reinhold's *Commentary on Peurbach's [Planetary] Theories.* I took much from Plutarch in my *Optical Part of Astronomy* in 1604 at Prague. It was not, however, because of the islands that Plutarch named in the Icelandic Ocean that I chose Iceland for the hypothesis of my *Dream.* One reason, among others, was that at that time there was on sale, in Prague, Lucian's book about the trip to the moon, translated into German by the son of Rollenhagen, along with stories of St. Brendan and St. Patrick's purgatory in the subterranean regions of Iceland's volcanic Mount Hekla. Since Plutarch, too, followed pagan theological opinion and put a purgatory for souls on the moon, it was most agreeable to me when I was about to set forth for the moon, to take off from Iceland. An even greater recommendation for this island was Tycho Brahe's account, concerning which see below. I was influenced also by recollection of the story I had read about the winter the Hollanders spent in icy Nova Sembla, which likewise afforded many astronomical exercises for my *Optical Part of Astronomy* in 1604.

ancients called Thule. Because of the recent death[3] of my
mother, Fiolxhilde,[4] I am free to write of something which I
have long wanted to write about. While she lived [5] she earnestly entreated me to remain silent. She used to say that
there are many wicked folk who despise the arts and interpret maliciously everything their own dull minds cannot
grasp.[6]

. . . . . .

[3] Because it is more natural for a son to disclose his mother's
secrets after she has gone than while she is still alive. I wished, too,
to hint that Science is born of untaught experience (or, to use
medical terms, the offspring Science has as its mother empirical
practice) and that so long as the mother, Ignorance, lives, it is
not safe for Science, the offspring, to divulge the hidden causes of
things; rather, age must be respected, a ripening of years must be
awaited, worn out by which, as if by old age, Ignorance will finally
die. The object of my *Dream* was to work out, through the example of the moon, an argument for the motion of the earth; or
rather, to overcome objections taken from the general opposition
of mankind. I believed that Ignorance was by then sufficiently extinct and erased from the memory of intelligent men. But the
spirit struggles in a chain of many links, strengthened by many
centuries, and the ancient mother is still alive in the Universities,
but living in such a way that death must seem better to her than
life.

[4] On the wall of living quarters I occupied by permission of
Martin Bachazek, Rector of Prague University, there hung a very
old map of Europe on which the word "Fiolx" was attached to
many places in Iceland. Whatever the significance of this word,
the rugged sound pleased me, and I added "hilde," a common
designation for females in the ancient tongue, as in Brunhilde,
Mathilde, Hildegarde, Hiltrud, and similar names.

[5] [Missing.]

[6] This happened to me on my recent trip, although not to me
alone but to a like-minded group of several. A theologian who
professed the Augsburg Confession attacked us with great zeal. He
was using scripture as a weapon with which to assail us. Finally,
aroused by our defending arguments, he called in a loud voice

They fasten harmful laws onto the human race;[7] and many, condemned by those laws,[8] have been swallowed by the

· · · · · ·

for sacred witnesses to his announcement that this doctrine "fights against all reason." Then I, breaking my obstinate silence (for I had been sitting there a mere auditor until that point) said: "It is this, no doubt, that drives your faction on, at least the ignorant. If your narrow mind could comprehend the usefulness and necessity and strength of this doctrine you would long ago have stopped taking opposition arguments from scripture and would instead have sought out some suitable explanation as you frequently do in other circumstances. Your reasoning is so weak that you do not see that there is some particle of reason on our side, too. Therefore, a teaching cannot be against all reason when it is not against the reasoning of astronomers and physicists. What one person does not understand may be understood by another person who has more knowledge of the subject."

[7] Each suffers his own injustice. The chief injustice to Copernicus' *Revolutions* is that persons who have no knowledge of astronomy (their strictures come not from the actual meaning of the book but from incorrect interpretation) are of the opinion that this work should not be read until the part about the motion of the earth is taken out, which is to say that it should not be read before being burned. Thinking that these persons should be refuted not with arguments but with laughter, I have written the following epigram:

> They castrated the poet
> Lest he copulate;
> He lived without testicles.
> O, Pythagoras,
> Whose thoughts survived chains:
> They allow you life
> After removing your brains.

[8] If I am not mistaken, the author of that insolent satire called *Ignatius, His Conclave*, got hold of a copy of this little work of mine; for he stings me by name at the very beginning. As he goes along he brings poor Copernicus before the tribunal of Pluto, to

abysses of Hekla.[9] My mother never told me my father's
· · · · · ·

which, if I am not mistaken, there is access through the abysses of
Hekla (notes 2, 5, 9). You, my friends, who have some knowledge
of my affairs, and know the cause of my last trip to Swabia, especially those of you who have previously seen this manuscript,
will judge that this writing and those affairs were ominous for me
and mine. Nor do I disagree. Ominous indeed is the infliction of a
deadly wound or the drinking of poison; and the spreading abroad
of this writing seems to have been equally ominous for my domes-
tic affairs. You would think a spark had fallen on dry wood; that
is, that my words had been taken up by dark minds which suspect
everything else of being dark. The first copy went from Prague to
Leipzig, thence was taken to Tübingen in 1611 by Baron von
Volckelsdorff and his tutors in morals and studies. Would you believe that in the barbershops (as if the name of my Fiolxhilde is
particularly ominous to people there by reason of their occupation) my little tale became the subject of conversation? Certainly
in the years immediately following, from that city and that house,
there issued slanderous talk about me, which, taken up by foolish
minds, became blazing rumor, fanned by ignorance and superstl-
tion. Unless I am mistaken, you will agree that my home might
have been without that plague of six years, and I without my recent year-long trip abroad, had I obeyed the instructions I dreamed
Fiolxhilde had given. It has pleased me, therefore, to avenge the
trouble my dream has caused me by publishing this work, which
will be another punishment for my adversaries.

[9] The history of Mount Hekla, the volcano, is known from maps
and geography books. In speaking of punishment, I was referring
to what I consider to be the legend of Empedocles as told by
Diogenes Laertius. According to the story, Empedocles ascended
Mount Etna to win honor from the gods after his death; he is
supposed to have let himself down into the crater, sacrificing himself alive to the flames. Actually, perhaps, in searching for the
causes of the everlasting fire, he fearlessly advanced to a point from
which he could not return, and the encrusted surface of the ashes
gave way beneath his feet; and, repenting too late of his curiosity,
he gave up his sorrowing spirit unwillingly, and without any care

name,[10] but she said he was a fisherman and that he died at the very old age of one hundred and fifty years (when I was three) after about seventy years of marriage.[11]

In my early childhood, my mother often would lead me by the hand or lift me onto her shoulders and carry me to Hekla's lower slopes.[12] These excursions were made especially around the time of the feast of St. John, when the sun, occupying the sky for the whole twenty-four hours, leaves no room for night.[13] Gathering various herbs there, she took them home and brewed them with elaborate ceremonies,[14] stuffing them

. . . . . .

for this renown. A similar fate befell Gaius Pliny, who, at the time of the eruption of Vesuvius, betook himself through deadly rains of ashes and dust for the purpose of gaining knowledge, and was suffocated when his nose and throat became filled with the sulphurous stench and ashes. According to legendary accounts, Homer, tormented by the riddle of the fisherman, and Aristotle, by the alternating current of the Euripus, lost their lives in the waves. Many others pay the penalty for love of knowledge by poverty and by provoking the hatred of the ignorant rich.

[10] I was making fun of the barbaric ways of the ignorant. If you would give Ignorance as the mother of Science, as I did above, but Reason as the father, it is natural that this father should be either unknown to the mother or concealed by her.

[11] In a historical description of Scotland and the Orkneys, written by Buchanan, mention is made of a fisherman who at the age of one hundred and fifty became father of several children by a young wife.

[12] Because higher up are snows and crags and at the summit flames from the subterranean regions, as histories bear witness.

[13] Because Iceland is situated near the polar circle. This I heard also from Tycho Brahe, who made this calculation on the basis of an account of a Bishop of Iceland.

[14] The studies of medicine and astronomy are related; they are from the same source, a desire for a knowledge of nature. Many superstitions are connected with the empirical knowledge of botany, however.

afterward into little goatskin sacks which she sold in the nearby harbor to sailors on ships, as charms.[15] In this way, she made a living.

Once, out of curiosity, I cut open a pouch unbeknownst to my mother who was in the act of selling it, and the herbs and patches of embroidered cloth[16] she had put inside it scattered all about. Angry with me for cheating her out of payment, she gave me to the captain in place of the little pouch so that she might keep the money. And he, setting out unexpectedly next day with a favorable wind, headed as if for Bergen in Norway.[17] After several days, a north wind came up;[18] blown off course between Norway and England, he headed for Denmark and traversed the strait, since he had a letter from a Bishop of Iceland [19] for Tycho Brahe, the Dane, who was living on the island of Hven. I became violently seasick from the motion and the unusual warmth of the breeze,[20]

. . . . . .

[15] The tradition, whether true or false, is common in geographical studies that the steersmen of ships sailing from Iceland call forth whatever wind they want by opening a bag of wind. If anyone should interpret this tradition in terms of the rhumbs of the compass rose, the magnetic pointer, and the guiding of the rudder, he would almost speak the truth. Since it is reckoned that there are thirty-two winds, whatever one of the sixteen in one hemisphere is blowing will speed the ship along the desired course of that hemisphere if a skillfull helmsman steers as the compass indicates; the winds of the opposite hemisphere can be rendered ineffective by traveling in different directions in what is called "tacking."

[16] A Bishop of Iceland told Tycho Brahe that Icelandic girls would weave at remarkable speed on bits of cloth, with needle and colored thread, words or phrases they heard in sermons at church.

[17] Having passed Scotland and the Orkneys in the upper ocean.

[18] Lacking the bag of winds, he could not escape the blasts of the north wind and reach his destination: Norway.

[19] From Tycho's account, as in note 13.

[20] Tycho Brahe writes to the Landgrave of Hesse that reindeer, a species of northern deer, do not thrive in Denmark because that

for I was in fact a youth of only fourteen. After the boat reached shore, the captain left me and the letter with an island fisherman,[21] and, having given me hope of his return, set sail.

Brahe, greatly delighted with the letter I gave him, began to ask me many questions[22] which I, unfamiliar with the language, did not understand except for a few words.[23] He therefore imposed upon his students, whom he supported in great numbers,[24] the task of talking with me frequently: so it came about, through this generosity of Brahe[25] and a few weeks' practice, that I spoke Danish fairly well. I was no less ready to talk than they were to question and I told them many new

. . . . . .

country, though cold, is warmer than Bothnia, Finland, and Lapland, the animal's native land. It is reasonable, then, to ascribe the same degree of cold to Iceland which is also situated near the arctic pole.

[21] This island's only inhabitants (it is barren, rocky, not very spacious) are about forty fishermen.

[22] It was the regular practice of this truly philosophical man to question, to learn, to have regard for such accounts, to ponder over them repeatedly, and to apply them to physical theorems.

[23] Although Teutonic, too, the dialect of Denmark is different from that of Iceland, which seems to be a colony of the Norwegians, who established dominion over it and over nearby Greenland one hundred years ago. From the account of a certain shipwrecked Venetian merchant, it is evident that the Orkneys also acquired the customs and the language of the Teutons two hundred years ago.

[24, 27] Seldom fewer than ten, sometimes as many as thirty. He trained them in the use of various instruments for observation of the stars, in drawing, in computations, in pyronomic works, and similar philosophical pursuits.

[25] He had control of a great hereditary fortune, which he spent freely on studies. Conquering all fatigue, he strove tirelessly toward goals commonly despaired of, as is demonstrated by the exceptionally accurate observations in which he engaged in battle with and overcame the very nature of human vision.

things about my homeland in return for the marvels they related to me.

Finally the captain of the ship that had brought me returned. But when he came to fetch me, he was sent away.
And I was very happy.[26]

The astronomical exercises pleased me greatly. Brahe and
his students passed whole nights with wonderful instruments
fixed on the moon and stars.[27] This reminded me of my
mother because she, too, used to commune constantly with
the moon.[28]

Thus by chance I, who came from very impoverished circumstances in a half-barbaric land, achieved an understanding
of the most divine science, which has prepared the way for
me to greater things.

After I had passed several years on the island, I was seized
with a desire to see my home again. I thought it would be no

. . . . . .

[26] It was among this man's pleasures sometimes to disappoint
with an unexpected rejection, right at the water's edge, those who
were on the point of leaving the island and had already been sent
off, and to detain them longer than they wished, unless a person
could fly.

[27] [See footnote 24.]

[28] I was at that time engaged in reading Martin Del Rio: *Investigations of Magic*. And there is a well-known line from Virgil:
"Incantations can even bring the moon down from the sky."

The region of the sky was appropriate above all others. For when
the moon is full for others, it often does not appear in Iceland.
Writers say that sorcery is common among peoples of the north,
and it is believable that those spirits of darkness take advantage of
the long nights there; for Iceland is truly hidden far to the north.
No doubt the leisure of the subdued light and the uninterrupted
nights is conducive to philosophy. The most illustrious Duke Julius
Friedrich of Württemberg, who on a memorable trip traveled also
through the north, states that he found the people there remarkably learned, and that they presented their philosophy to foreigners
with an elegance rare among us.

difficult matter for me, with the knowledge that I had acquired, to attain some high office among my rude countrymen. Therefore, after obtaining my patron's approval for my departure, I left him and went to Copenhagen. There some travelers who wanted to learn the language and the region took me into their company; with them I returned to my native land five years after I had left it.

I was most happy to find my mother still alive, still engaged in the same pursuits as before. And the sight of me, unharmed and thriving, brought an end to her prolonged regret at having rashly sent her son away. Autumn[29] was approaching, and the nights were lengthening toward the time in Christ's natal month when the sun appears only briefly at midday before straightaway hiding itself again.[30] During this holiday from her labors, my mother clung to me continually; wherever I betook myself with my letters of recommendation, she did not leave my side. She kept asking me, now about the lands which I had visited, and now about the sky. She delighted in the knowledge I had acquired about the sky. She compared my reports of it with discoveries she herself had made about it.[31] She said she was ready to die since her knowledge, her only possession, would [now] be left to her son and heir.[32]

By nature eager for knowledge, I asked about her arts and what teachers she had had in a land so far removed from others. One day when there was time for conversation she told me everything from the beginning, much as follows:

Duracotus, my son, she said, provision has been made not

. . . . . .

[29] I considered this time of year the most suitable for a voyage from that port in the kingdom of Denmark to Iceland.

[30] This follows from note 13, in accord with spherical doctrine.

[31] See note 28.

[32]
        Chariot-drivers dream of chariots,
            Judges dream of the legal fight;
        What you seek by daylight,
            You find at night.

only for the regions you have visited, but for our land, too. For although we have cold and darkness and other inconveniences, which now at last I am aware of when I learn from you the delights of other regions, we are nonetheless well endowed with natural ability,[33] and there are present among us very wise spirits[34] who, finding the noise of the multitude and the excessive light of other regions irksome, seek the solace of our shadows and communicate with us as friends. Nine[35] of these spirits are especially worthy of note. One,[36] particularly friendly to me, most gentle and purest of all,[37] is called forth by twenty-one characters.[38]

. . . . . .

[33] The above-mentioned Bishop told Tycho Brahe that the Icelanders are exceedingly talented.

[34] These spirits are the sciences, which reveal the causes of things. This allegory was suggested to me by the Greek word *daimon*, which is derived from *daiein*, that is, "to know," as if it were *daêmon*. With this substitution in mind, read note 28 from "No doubt."

[35] The exact reason for this number escapes me. Or was I referring, perhaps, to the nine Muses, because these, too, are goddesses to the heathen, and thus to me are spirits? Or did I adapt these (the sciences) to the number: (1) Metaphysics, (2) Natural Science, (3) Ethics, (4) Astronomy, (5) Astrology, (6) Optics, (7) Music, (8) Geometry, (9) Arithmetic?

[36] I am certain that here I had in mind either Urania among the Muses or Astronomy among the sciences. For apart from the fact that because of the cold they lack many of the necessities of life, I should say that the inhabitants of the north are more suited than others to astronomy because the divisions of day and night (which are conducive to the study of astronomy) are greater among them.

[37] If this statement concerns the Muses, the charge of vanity is made against the others. But if about the sciences, natural science also teaches poisons and if injudiciously pursued it engenders even alchemy; metaphysics has preposterously inflated aspirations, and confuses generally accepted doctrines with excessive and troublesome subtleties; ethics recommends high-mindedness, which is

With his help I am transported in a moment of time to any foreign shore[39] I choose, or, if the distance is too great for me,[40] I learn as much by asking him as I would by going there myself.[41] Most of what you have seen, or learned

. . . . . .

not advantageous to all; astrology supports superstitions; optics deceives; music attends upon love; geometry serves unjust rule; arithmetic serves avarice. A better meaning would be that although all the sciences are gentle and harmless in themselves (and on that account they are not those wicked and good-for-nothing spirits with whom witches and fortune-tellers have dealings, who give irrefutable proof of their cruelty in the identity of their patron, Porphyry), this is especially true of astronomy because of the very nature of its subject matter.

[38] In seeking my reason for this number I got no further than the discovery that this is the number of letters or characters in the words *Astronomia Copernicana*. It is also the number of possible conjunctions between pairs of planets, of which there are seven. It amuses me, too, that such is the number of possible throws of dice. For 21 is a triangular number, with a base of 6. The allegory of "calling forth" was taken from Del Rio and from sorcery; nevertheless, there is a philological meaning: "to be called forth" is "to be stated as a proposition."

[39] Olaus and others mention this also about the Finns, a northern people, and about their neighbors, the Lapps. I applied it to the doctrine of natural days, zones, and regions, and to the experience of the Hollanders in the frozen sea, where they found everything arranged just as we astronomers here at a distance have known and taught it to be for many centuries.

[40] Consider now as assumed with respect to the heavens those same assumptions that were made before with regard to earthly regions approachable by men.

[41] There is a popular jest: "I'll believe it rather than go into it myself." Many ask whether we astronomers have just fallen down from the sky. To them, Galileo's *Star Messenger* replies according to his own notion; but the judgment of reason is stronger, an unexceptionable witness, as the Hollanders learned from experience in those winter quarters of theirs. See note 39.

from conversations, or drawn from books, he has already reported to me, just as you have. I should like you to go with me now to a region he has talked to me about many times, for what he has told me is indeed marvelous. She called it Levania.[42]

Straightaway I agreed that she should summon her teacher. It was now spring; the moon, becoming crescent, began to shine as soon as the sun dropped below the horizon, and it was joined by the planet Saturn, in the sign of Taurus,[43] just

. . . . . .

[42] The moon in Hebrew is *Lebana*, or *Levana*. I could have called it Selenitis. But Hebrew words, which are less often heard by us, are recommended for occult arts by the greater aura of superstition attached to them.

[43] See how the very necessity of my suppositions has again cast me up on the same shore that Plutarch chooses; for he, too, mentions the return of Saturn into Taurus. But I arrived at my choice as follows. I adopted the manner of the astrologers and established the sun and the moon in their own respective dignities. The sun was in its house in Leo, however, before Duracotus returned home. If this had not been the case, the sun, to be sure, could have been placed in that sign, and the moon, both waning and crescent, could have been placed in Cancer, likewise its house. But this would have been unsuitable because, night being more appropriate for these events, the sun had to be below the eastern horizon. On account of these inconveniences, I deleted the sign of Cancer and deserted the year in which Saturn was in Cancer—that is, 1593, the year when you wrote the disputation on the moon, Besold. I still find these erasures in my first copy. Figuring that my Duracotus had spent the winter in his native land, I chose March, instead, with the sun at equinox, a good astronomical sign, and on the porch of Aries, its own exaltation. If, therefore, the moon was to appear horned, that is, in a sign near the sun, it could have no other dignity than exaltation in Taurus. In order that it might be seen next to other stars, I put the sun below the western horizon, at the beginning of the night. This was especially desirable in the signs on the long descensions, for the whole body of the moon is

after sunset. My mother withdrew from me to a nearby cross-roads,[44] and after crying aloud a few words[45] in which she set

. . . . . .

then quite visible within the embrace of its shining crescent, as I said in my *Optics* and in my *Conversation with the Star Messenger* of Galileo and finally in my *Epitome*. That Saturn, too, might be in conjunction with the moon (a conjunction astrologers consider a sign of occult arts), Saturn also had to be placed in Taurus. The time turned out to be the period of Tycho's most frequent observations: in the year 1589 11/21 March in the evening, when conjunction likewise occurred with the constellation of the Pleiades. Anyone born when these stars are near the moon is likely to be imaginatively endowed, as I granted in *Harmonics*. Even astronomical reasons commend this constellation for observation of the horned moon. My *Optics* (Chapter XI, page 347) has an observation of this sort for the year 1589, on April 8 and July 27.

[44, 46, 47] This, too, is a magic ceremony to which there is a correspondence in the method of teaching astronomy, in that it is by no means an offhand matter, one for a public teacher. But every useful reply requires repose, attentiveness, and carefully selected words. Whenever men or women came to watch me in a particular observation which I performed frequently in Prague in those years, I would first remove myself from them to a nearby corner of the house which I had chosen for this activity. There, after shutting out the daylight, I would fashion a small window out of a tiny aperture, and put a white covering on the wall opposite. Having done all this, I would summon the spectators. These were my ceremonies, my rituals. Do you want characters, too? On a black tablet I would write with a piece of chalk, in big letters, whatever seemed suited to the spectators. I would write the letters backwards (more magic!) as Hebrew is written. When the slate was hung upside down in the sun outside the house, the letters I had written would be reflected right side up against the white cloth on the wall within. If a breeze should agitate the slate, the letters gave the impression that the wall inside was moving slightly.

[45] Those aforementioned spectators who are left will see, when they put their minds to it, what that crossroad in my house was. But here is understood the astronomical crossroads in the hypothesized celestial arrangement, and it is double: one is where

forth her desire,[46] and then, performing some ceremonies, she returned,[47] right hand outstretched, palm upward, and sat down beside me.[48] Scarcely had we got our heads covered with our robes[49] (as was the agreement) when there arose a hollow, indistinct voice,[50] speaking in Icelandic to this effect:

. . . . . .

the sun is fixed at the equinoctial point, at which the equator and the ecliptic path of the sun cross each other. In Brahe's manuscript is an observation he made of the altitude of the sun at equinox on that very day. The other crossroad is the descending node of the moon or dragon's tail, which was then at the end of Aquarius. The astronomer must pay attention to this node to tell when the moon has reached its limits. It was indeed then in its southern limit at the end of Taurus, a position that invites observation of the latitude of its limits.

[46] [See footnote 44.]

[47] [See footnote 44.]

[48] I used also to throw in these very games, which were especially pleasing to my spectators because they knew that they were games.

[49] With this very ritual (how magically magic!) we had observed an eclipse of the sun a little before I got the idea for this book, that is 1605 2/12 October. You who were there remember, you envoys from the Palatine of Neuchâtel. On the terrace of the charming house in the Emperor's garden we had no dark room, so we cut off the daylight by enveloping our heads in our garments.

[50] I do not consider it impossible, by means of various instruments, to produce both individual vowels and consonants in imitation of the human voice. Nevertheless, whatever comes forth will be closer to a rumbling and mumbling than to a lifelike voice. Yet even in this device, I believe, are traps for the superstitious and the credulous, so that sometimes they will think that demons are talking to them, when art is imitating magic tricks. Nevertheless, whatever the truth of this may be, I consider it more correct to believe those who affirm it properly than to deny it when I can rely upon no personal experience.

There comes to my mind the delightful memory of the renowned Matthias Seiffart, a fellow initiate left by Tycho Brahe to

THE DAEMON[51] FROM LEVANIA[52]

Fifty thousand German miles[53] up in the air the island of Levania lies.[54] The road to it from here, or from it to the earth, seldom lies open[55] for us.[56] Indeed, when it does, it is

. . . . . .

his heirs, who spent three months in computing, from the teachings of Tycho, the ephemerides of the moon for one year. Now his voice was not dissimilar to what I have described, and he suffered also from melancholy and from a mental disorder in which there was no place for recreation, and which eventuated in a fatal dropsy.

   [51] The knowledge of the phenomena of the stars, from *daiein*: "to know."

   [52] Derived from the word for moon, to which the eyes were fictionally transported.

   [53] To each degree of the great circle of the earth are ascribed fifteen German miles. At Rome, for example, the altitude of the pole is 41° 50′; at Nuremberg, on almost the same meridian, 49° 26′; therefore, it is 114 miles from Nuremberg to Rome, and to the bank of the Danube it is 100 miles. At Rostock, the altitude of the pole is 54° 10′; therefore, from Nuremberg to Rostock is 71 miles. So at Linz, the altitude of the pole is 48° 16′ and at Prague 50° 6′—a difference of 1° 50′, reckoned as 26 miles. If 1° consists of 15 miles, there will be 860 miles in the radius of the circle which is the circumference of the earth. I demonstrate in my *Hipparchus*, and I deduce a priori in *Epitome of Copernican Astronomy*, that the moon at apogee is about 59 earth radii away from the earth; 860 multiplied by 59 makes 50,740 miles.

   [54] It is not "located" but rather, if we have regard for the metaphor of an island, it floats. But the expression here had to be in terms of its visual impression. For a person who was on the moon would certainly think that the moon was fixed in one place.

   [55] Here reasoning of physics is combined with a jest as to why eclipses of the sun and moon bring about so many ills. Evil spirits are certainly called powers of darkness and air. It might be believed, therefore, that they have been condemned and, so to speak,

easy for us, but for men the passage is exceedingly difficult, and made at grave risk to life.[57] No inactive persons are accepted into our company; no fat ones; no pleasure-loving ones;[58] we choose only those who have spent their lives on horseback, or have shipped often to the Indies and are accustomed to subsisting on hardtack, garlic, dried fish, and such unpalatable fare.[59] Especially suited are dried-up old crones,[60]

. . . . . .

relegated to the regions in the cone of the earth's shadow. When this cone of shadow touches the moon, the daemons use the cone of shadow as a ladder to invade the moon in great swarms. And when the cone of the moon's shadow touches the earth in a total eclipse of the sun, the daemons return to earth through the cone, as below in note 86. These occasions are rare. Insofar as a daemon is here understood to be the science of astronomy, the assertion that there is no other way for the mind to go to the moon than by means of the shadow of the earth and by whatever other means depend on this shadow is a serious one. See "Shadow Measurement," a part of my *Hipparchus*.

[56] If we pursue the allegory, it is easy to arrive at a knowledge of heavenly affairs with the help of shadow measurement. But if we are pondering the nature of bodies and of spirits, the reasoning behind what is asserted here is again obvious. I indulge in this little joke with my thoughts intent on physics while I cast satiric little darts in all directions at self-confident spectators.

[57] Understand this as a matter of physics, if a body bearing its own weight should be carried aloft twelve thousand miles in an hour. Add the lack of air (see note 71), which is as fatal to men as lack of water is to fish. For among the doctrines which I consider established by the most eminent physicists is this: the outer reaches of the air terminate at the peaks of the highest mountains or even lower down.

[58] Tycho Brahe in his writings frequently abuses men of this sort who boast of their learning while he does the work and keeps the night-long vigils.

[59] I had reference here to an epigram with which I made fun of the bodily type of Maestlin, who was then my teacher:

who since childhood have ridden over great stretches of the
earth at night in tattered cloaks on goats or pitchforks. No
Germans are suitable, but we do not despise the lean hard
bodies of the Spaniards.[61]

The whole of the journey is accomplished in the space of
four hours at most.[62] For, busy as we always are, we agree not

. . . . . .

> The slighter the flesh
> On a person's bones,
> The quicker he flies
> To heavenly homes.

The subtlety of a penetrating intellect is well known. But for-
give me, you who are stung.

[60] Behold Aulis and the alliance that destroyed Troy! Such a desire had I to jest and to argue jestingly! But, I say, if what most
tribunals say about witches is true—that they are carried through
the air—even this will perhaps be possible, that a body separated
from the earth might be carried to the moon.

[61] This, too, is part of the paraphernalia of humor, that while
you figure on winning the applause of one while another is listening, you wound both one and the other. Nevertheless, just as the
Germans have a reputation for corpulence and gluttony, so the
Spaniards are renowned for their genius and judgment and frugality. Therefore, in precise sciences, of which astronomy is one (and
especially this lunar astronomy, which is situated in an unusual
position, if anyone were looking out from the moon), if a German
and a Spaniard should strive equally, the Spaniard would outstrip
the German by far. Therefore, I could foresee that this work of
mine would be a source of mockery for the Germans but that the
Spaniards would take some account of it.

[62] The duration of a central eclipse of the moon, from beginning
to end, exceeds this by a few minutes when the two bodies are at
apogee. For the parallax of the sun is $0'$ $59''$ and that of the moon
$58'$ $22''$, a total of $59'$ $21''$; since the radius of the sun is $15'$ $0''$, the
radius of the shadow is $44'$ $21''$. When the $15'$ $0''$ radius of the
moon is added to this, we return to the total of $59'$ $21''$. But the
true hourly schedule of the moon is $29'$ $44''$, that of the sun is

. . . . . .

2′ 23″, and the difference between the sun and the moon is 27′ 21″. Doubling this for the two hours, and subtracting, leaves 4′ 39″. Of this, 4′ 33″ 30‴ are accomplished in ten minutes and the remaining 5″ 30‴ in twelve seconds. Therefore, the entire duration is 4 hours, 20 minutes, 25 seconds. This great length is very uncommon, however; therefore, if a body is carried from the earth to the moon it must either circulate aloft for several days in the cone of the earth's shadow (in order that it be on hand at the moment of the moon's entry to this cone) or, if this is quite contrary and opposed to the nature of the body, the whole trip from earth to the moon will be made in that very short time during which the moon is in the cone of shadow. The theory of magnetism provides us with another reason. The moon is a body related to earthly matter. Plutarch asserts this in many ways through one of the speakers in his book, *The Face on the Moon*, which is appended here. [*Note:* It is not included in this translation.] And Aristotle's Arab interpreters forced even him to this side. Unless I am mistaken, they urge upon us that passage in Book II of *The Heavens* concerning which see the preface to Book IV of my *Epitome of Copernican Astronomy*, chapter 12. But the clearest evidence of relationship between earth and the moon is the ebb and flow of the seas, concerning which see my introduction to the *New Astronomy:* [*A Description of Celestial Physics, with*] *Comments on the Motions of the Planet Mars*. When the moon is at zenith over the Atlantic Ocean, the so-called southern ocean, the eastern ocean, the Indian Ocean, it draws the waters which are spread out all over the globe; and it comes about, through this attraction, that the waters hastening from everywhere to the vast area which is perpendicular to the moon and is not cut off by continents, lay bare the shore. In the meantime, while they are on their journey, the moon departs from the zenith of one ocean; deserted by the drawing force, the waters dashing against the western shore fall back and pour onto the eastern shore. In the last chapter of Book IV of *Harmonics*, I discussed a further cause of the ebb and flow of the sea, which is indeed connected with this. But what I touch upon here suffices for the present purpose. For if the daemons exist nowhere but in the cone of shadow, and if they are imagined to snatch a body aloft toward the vertex of the cone, certainly, to leave[63] before the eastern edge of the moon begins to go into eclipse. If the moon should shine forth full while we were still en route, our departure would be in vain. On such a headlong dash, we can take few human companions—only those who are most respectful of us.[64] We congregate in force and seize a man of this sort; all together lifting him from beneath, we carry him aloft.[65] The first getting into motion is very hard on him,[66] for he is twisted and turned just as if, shot from a cannon, he were sailing across mountains and

. . . . . .

unless the moon is traveling through the cone at the same time, they will be alone, unaided, they will labor and sweat and, of course, grow faint. If, however, they undertake the task when the moon is right, the moon, by its presence in the shadow, will aid them in their attempts by the magnetic attraction of her related body. See below, note 78.

[63] Another reason for allowing no more than the length of an eclipse for this transportation is derived not from the nature of the body but from the disposition of those being transported.

[64] This whole sentence has to do with the allegory. Since notable and extensive eclipses are rare, and opportunities for observing them are rare, the science of astronomy (one of the spirits) does not become commonly known by means of eclipses. But there are philosophers who cherish all philosophical sciences (the family of these spirits, that is) exceedingly. These, I say, lie in wait for lunar eclipses and, using them as a ladder, dare the ascent to the moon: that is, attempt an investigation into the nature and courses of heavenly bodies. Only the smallest part of the human race applies itself to philosophy, and in the ranks of the philosophers scarcely one or two endeavor to extend the boundaries of astronomy.

[65] I here return to a contemplation of the nature of the bodies, employing a fiction.

[66] I define gravity as a power similar to magnetic power—a mutual attraction. The attractive force is greater in the case of two bodies that are near to each other than it is in the case of bodies that are far apart. Therefore, bodies more strongly resist separation one from the other when they are still close together.

seas.[67] Therefore, he must be put to sleep beforehand, with narcotics and opiates,[68] and he must be arranged, limb by limb,[69] so that the shock will be distributed over the individual members, lest the upper part of his body be carried away from the fundament, or his head be torn from his shoulders. Then comes a new difficulty: terriffic cold [70] and difficulty in breath-

· · · · · ·

[67] The force is not very powerful when the body being pushed moves easily. A ball of lead is more violently jarred than one of stone because there is more weight in the former, hence more resistance. Since, therefore, the bodies are heavy, they will resist motion, hence the force of such a quick thrust will be very violent.

[68] I have been mindful of the intensity of the suffering, anyway. Let someone else have care for the safety of the traveler, lest he be ground into little bits, no matter whether sleeping or waking.

[69] Those parts of the body which are crowded together nearest to the driving force, being pressed by the weight of parts lying on top of them, suffer most.

[70] Our bodies are warmed by continuous evaporation from the bowels of the earth, which falls either in rain, or in the night (when the warm rays of the sun are absent) in dew or frost. The skin, when deprived of this warm vapor from without, begins to roughen. Also, when vapor emanating from the body loses the heat by force of which it was exuded, it coagulates and becomes cold matter. And, in the process of coagulation, it acquires a motion toward the body, which is its source, and, attacking the body, chills it. Finally, the ethereal air, deprived of the sun's rays, is cold by reason of this lack of heat. As the air is very thin, it absorbs only a very slight degree of cold as long as the air is motionless. But motion, when it is introduced, by the very fact of its existence, brings to the air a certain density, so that the more violent the impact of it on a body or the impact of a flying body on it, the denser it becomes and the more penetrating by reason of its subtlety and hence likewise the colder. Cold becomes an active quality by means of the thickening of matter; when matter is not yet condensed, I recognize it as only negatively cold. I leave the transition from a negative to a positive form to others to explain.

ing.[71] The former we counter with our innate power,[72] the latter by means of moistened sponges applied to the nostrils.[73] When the first part of the trip is accomplished, the carrying becomes easier.[74] Then we entrust the bodies to the empty air and withdraw our hands.[75] The bodies roll themselves together into balls, as spiders do, and we carry them almost by means of our will alone.[76] Finally, the corporeal mass heads of its own accord for the appointed place.[77] But this spontaneous

. . . . . .

See the speculation on this subject in my *Optics*, where I set it forth by means of a comparison of light and of black color; where you see me laboring, help me dig out the causes.

[71] See note 57, above.

[72] This is just for form's sake. I do not know whether it is a good thing to jest in a serious matter. Even the allegory is frozen. The daemon, called astronomy, by the innate strength of his burning passion for speculation, quite feebly supplies the necessities of life.

[73] I could not pass by Aristotle's very appropriate story about the philosophers struggling toward Mount Olympus in Asia for the sake of observation.

[74] That is, when the body was separated from the orbit of the earth's magnetic attraction by so great a distance that the magnetic power of the moon preponderated.

[75] When the magnetic powers of the moon and earth are in contrary attraction, it is just as if the body were drawn in no direction at all. The body, therefore, as a whole, itself then attracts its members, as the parts are less than the whole.

[76] Not simply by our will. There is need of some force also. For every body by reason of its own matter has a certain inertia in regard to motion which provides repose to the body in every place in which the body is placed in a position beyond powers of attraction. Whoever would move this body from its place must overcome this force, or, rather, this inertia.

[77] When, of course, on account of its propinquity, the magnetic attraction of the orbit of the moon predominates. Take any mass of earth you wish that is equal to the mass of the moon. It will draw equally powerfully a body placed between the two globes in

motion is not very useful because it occurs too slowly.[78] Therefore, as I have said, we hurry the bodies along by willpower,
and then we precede them, lest some damage be inflicted by
a very hard impact on the moon. When the men awaken,
they usually complain about unspeakable weariness in all their
limbs; but they later entirely recover from this so that they
can walk.[79]

Many further difficulties arise which would be too numerous to recount. Absolutely no harm befalls us. For we stay in
close array within the shadow of the earth,[80] however long it
is, until the shadow reaches Levania,[81] when we disembark, as
if from a ship. There we betake ourselves hurriedly to caves
and to shadowy regions,[82] in order that the sun, coming out

. . . . . .

such a way that the relation of its distance from both is the same
as that of the bodies from each other; the body will dangle im-
mobile, as the opposing attractions cancel each other out. And if a
body should be 58 1/59 earth radii distant from earth and 58/59
earth radii distant from the moon, that body also would be held
motionless. But as soon as the body approaches slightly closer to
the moon, the body will follow the gravitational pull of the
moon, whose force has become predominant because of propinquity.

[78] At first not very [useful], but near the moon extremely so.
Moreover, it is not very useful to those who are not really making
an effort, but it is very helpful to those struggling to raise the body
even when the [power of the] earth still predominates. See note 62,
above.

[79] See notes 67, 68, 69. Let the traveler see to it that he arrives
in such an unharmed condition that he can awaken. The allegory
here furnishes an approved medicine for those who are bound by
a vow of virginity against natural motions: intent, constant, and
ardent observation.

[80] Which, as I said in note 62 above, is quite contrary and opposed to the nature of the body, even impossible.

[81] I am greatly pleased to find almost the same words in Plutarch.

[82] Because they are called spirits of darkness, as in note 55 above.

into the open a little later, may not overwhelm us and drive us from our intended lodging and force us to follow the departing shadow.[83] We are granted a truce for exercising our talents as the spirit moves us; we confer with the daemons of this province, and entering into an alliance with them, as soon as a place begins to lack the sun, we join forces and spread out into the shadow. If the shadow of the moon strikes the earth with its sharp point,[84] as generally happens,[85] we, too, attack the earth with our allied armies. This can be done at no other time than when men see the sun go into eclipse; it is for this reason that men so greatly fear eclipses of the sun.[86]

So much for the trip to Levania. Now let me tell you about the place itself, starting, as the geographers do, with those things which are determined by the heavens.

Although all of Levania has the same view of the fixed stars as we have,[87] it has, nevertheless, such a different view

. . . . . .

But the allegory compares observation of eclipses to a journey through the shadow, political activities to the sun, solitude and scholastic shadows to the shadowy caverns of the moon, prolonged speculation concentrated on observations of eclipses to the tarrying in caves. I had a house in Prague in which there was no corner more suitable for observation of the sun's diameter than an underground beer cellar, from the floor of which I used to point an astronomical tube (described in my *Optics*) through an opening aloft toward the midday sun on the days of the solstices. But this part of the allegory is elaborated in note 83.

[83] [Missing.]

[84] Which happens on the seventh or eighth day after an eclipse of the moon, as explained below.

[85] Total eclipses of the sun are more common than total eclipses of the moon.

[86] This was set forth more explicitly in note 55 above. But if you continue according to the allegory, you will elicit predictions of solar eclipses from speculation depending on observation of lunar eclipses.

[87] For if, as I deduced a priori in *Copernican Astronomy*, Book

of the motions and sizes of the planets that it must surely
have a wholly different system of astronomy.

Just as our geographers divide the earth's circle into five
zones with respect to heavenly phenomena, so Levania consists of two hemispheres:[88] one Subvolva, the other Privolva.[89] The former always enjoys Volva, which is to it
what our moon is to us; while the latter is forever deprived of
the sight of Volva.[90] [See footnote 89.] A circle dividing the
hemispheres in the manner of our solstitial colure [i.e., a cir-

. . . . . .

IV, the radius of the sun is to the radius of Saturn's orbit as this
radius is to the radius of the fixed stars, then it follows that the
radius of Saturn's orbit is scarcely 1/2000th of the radius of the
sphere of the fixed stars; the radius of the earth-moon system is
scarcely 1/20,000th; and the radius of the moon is 1/59th of the
earth-moon radius. Therefore, the distance from the earth to the
fixed stars varies by no more than 1/10,000th, and the digressions of
the moon add about 1/30th of this small variation. Thus the
entire digression of the moon from the fixed stars is quite imperceptible.

[88] Because the moon always turns the same spots to us earth
inhabitants, we know that the moon goes around the earth as if
attached by a thread, and that the upper half of the moon never
has a view of the earth but the lower hemisphere always has.

[89, 90] What we earth-dwellers call earth, I thought people on the
moon would fancy as Volva. Among us, the nocturnal luminary,
from its white color, is called *Lebana* in Hebrew, and in the Etrus-
can dialect (derived, I believe, from Carthaginian) it is called
Luna, and in Greek *Selene*, which means a white gleam (since it
looks like this to us who are situated on earth), so it is reasonable
to assume that moon-dwellers, who would see our earth as a
sort of moon, would give earth a name that has some relation to
its appearance. Our planet would appear to them to be in perpetual
revolution around its own unmoving axis; they have evidence of
this in the diversity of the spots, as will be shown below. Let this
revolving thing be called Volva, therefore, and let those who see
Volva be called Subvolvans and let those who are without sight
of Volva be called Privolvans.

cle on the celestial sphere which passes through the poles
and solstices] passes through the cosmic poles and is called
the divider [because it divides Subvolva from Privolva, Kepler
so named the circumference of the lunar hemisphere that is
visible to earth].[91]

I shall first set forth the features that are common to both
hemispheres. All of Levania experiences the same succession
of days and nights as we do,[92] but in the year as a whole the

· · · · · ·

[91] We on earth consider as the cosmic poles those two opposite
points in the sphere of the fixed stars to which the axis of the
earth would reach if it were extended at both ends. So far as the
appearance of first motion is concerned, we see these two points
as motionless. These points are not regarded as the cosmic poles by
the lunar people, for to them the starry sky does not appear to
revolve around them in that brief space of time which to us is
twenty-four hours. Instead, the axis of the lunar body, if extended
to the plane of the ecliptic, meets points in the sphere of the fixed
stars near the poles of the ecliptic. These points are the cosmic
poles to the moon-dwellers, because the sphere of the fixed stars
is perceived on the moon as revolving around this axis in the space
of time which we call a month. The globe of the moon really revolves on this axis and the two extremities of it, which are, so to
speak, motionless in place. Although the moon and this axis of
lunar motion go around the earth in the space of a month, the
moon meanwhile remains parallel to itself in each of its respective
positions. Since the size of the moon's orbit is imperceptible in
comparison to the size of the sphere of the fixed stars, the moon's
axis throughout one revolution points to almost the same spot
among the stars. That the divider circle passes through the poles
of the lunar globe is apparent from the fact that the same spots
of the moon are turned toward the earth for the whole period of
the monthly circuit. The more immovable we perceive the moon
as being, the more surely does it revolve around the above-mentioned points.

[92] Because the globe of the moon goes around the earth in such
a way that the same hemisphere is always turned toward the

people there lack our annual variation.[93] For throughout
Levania the days are almost equal to the nights, except that
each day is by the same fixed amount shorter than the night
for Privolva and longer for Subvolva.[94] What variation there is

· · · · · ·

earth, you can call this hemisphere the front of the moon. Obviously, when the moon is between the sun and the earth and appears to us new, or with slender horn, then the moon's back is to
the sun and its front is away from the sun. When the moon is
full for us, that is, when we are interposed between it and the
sun, its back, averted from the sun, is turned to the fixed stars and
its front is turned both to the sun and to the earth. In the entire
universe, the sun's presence makes day and its absence night;
therefore, the front and the back of the moon have their day and
night, but not such a brief day-and-night as we. What is to earth-
dwellers a whole month is consumed for the moon-dwellers within
the length of a single day and night.

[93] On earth the length of day and night (except at the time of
the equinox) varies because the cosmic poles appear to be far
distant from the poles of the ecliptic. The moon-dwellers do not
have our cosmic poles, but, on the contrary, have cosmic poles
near the poles of the ecliptic. If there is any variation [in the
length of their days and nights] as a result of the slight distance
[of their cosmic poles] from the poles of the ecliptic, this is certainly not very perceptible and not to be compared with the variation we have. Therefore, they have almost a perpetual equinox
throughout the whole globe, just as our earthly equinoctial day is
experienced by the whole earth.

[94] In my *Ephemerides*, the phase of the sickle-shaped moon is
separated from quadrature of moon and sun by 2 hours and 10
minutes at most. This is so because the ratio of the orbit of the
moon to the orbit of the sun (or of the earth) is 1:59—at apogee,
that is. But day dawns over the midregions of Subvolva when the
globe of the moon is situated in the circle of the illumination
of the moon which forms a phase—not when the center of the
globe of the moon is in quadrature with the sun. Since in both
phases the moon is closer to the sun than the earth is, that part
of the orbit of the moon that is drawn from these limits on the

in the course of eight years will be told below. Now at both
poles the sun, balancing between day and night, circles
around the mountains of the horizon,[95] half of it hidden
and half of it shining. For Levania seems to its inhabitants to
be stationary, while the stars go around it, just as the earth
seems to us to be stationary.[96] A night and a day, joined together, equal one of our months, for, in fact, when the sun
is about to rise in the morning, almost a whole sign of the
zodiac is visible that was not visible the day before.[97] And just

. . . . . .

outside around the earth is longer than the part that extends on
the inside between the earth and the sun. And that outside part
measures day for the midregions of Subvolva and night for the
midregions of Privolva. Therefore, the day of those on Subvolva
and the night of the Privolvans includes about four of our hours
in addition to half of our month; these same four hours are lacking from the Subvolvan night and the Privolvan day. The things
that have been said here about the midregions of the hemispheres
pertain also to the regions toward the east and west, with just this
difference: that by however many degrees of longitude any place
is removed from the middle, by the same number of degrees does
the globe of the moon also go beyond that location in which day
begins for the midregions.

[95] Just as, on the day of the equinox, it must happen at the very
pole on the globe of earth.

[96] Here is the hypothesis of the whole dream: that is, an argument for the motion of earth, or rather a refutation of arguments
constructed, on the basis of perception, against the motion of the
earth.

[97] If day is determined by the sun's presence, and if the sun is
present where we see places illuminated by its light, certainly for
the whole of fifteen days the same spots of the moon, in the
middle of its body, are experiencing day. For these places are seen
by us to be constantly illuminated, with no night intervening. The
full moon affords us a view of the spots while we remain in one
place for more than sixteen hours. Besides, when the moon is
hidden from us under the horizon, it appears to others because of

as for us in one year there are 365 revolutions of the sun and 366 of the sphere of fixed stars, or to be more precise in four years there are 1,461 revolutions of the sun but 1,465 revolutions of the fixed stars, so for those on Levania the sun revolves 12 times in one year and the sphere of the fixed stars 13 times in one year, or to be more precise in eight years the sun goes around 99 times, the sphere of the fixed stars 107 times. But a cycle of nineteen years [the so-called Metonic Cycle] is more familiar to them. For in that number of years the sun rises 235 times but the fixed stars rise 254 times.[98]

The sun rises in the centermost part of the region of Subvolva when the last quarter of the moon appears to us, but it shines in the centermost part of Privolva when the first quarter of the moon appears to us. These things that I say in regard to the middle regions are to be understood also with respect to the whole of the semicircles drawn through the poles and the mid-regions at right angles to the divider; you could call these the semicircles of the mediavolva [the central meridian of Subvolva].[99]

. . . . . .

the roundness of the earth. Therefore, their day is fifteen times as long as our day-night. And thus it follows that their night is fourteen or more times greater than ours.

[98] If we on earth—not people in general, but astronomers—count 99 months in eight years, or 235 months in nineteen years (although natural lunations are not as immediately important in our affairs as are nights and days), what can we expect of the lunar people (whose existence we are assuming) other than that they should observe the same numbers, if there is any creature capable of numbering, since they have no other day? To them the sign of an exact period of nineteen years is the rising of the same stars in exactly the same arrangement as before.

[99] Mediavolva is designated in the manner of our meridians. But earth has many meridians, while the moon has only one mediavolva, which in fact passes through only the two exactly

Now there is a circle halfway between the poles that performs the function of our earthly equator, by which name it shall accordingly be called; it cuts in half both the divider and the mediavolva at opposite points; and the sun daily passes almost directly through the zenith of whatever places lie on this circle, and it passes exactly over it on two opposite days of the year, at noon. In the regions that lie toward either pole, the sun at midday declines from the zenith.[100]

They have on Levania also some alternation of summer

. . . . . .

opposite points of the hemispheres which are named from Volva. However, the mediavolva is not a replacement for our meridians, for the lunar people have their own meridians, which go through the poles and the zeniths of points and are related to the mediavolva. Our earthly meridians have no natural point of origin, but theirs do: that is, the midpoint of the mediavolva, the place where the sun and Volva fall simultaneously at the same moment; on other lunar meridians, this does not happen at the same moment but at different moments.

[100] Since the moon is a globe, all heavy bodies on the moon will seek its center, and bodies will press upon the surface of the globe at right angles, and they will regard as their zenith among the fixed stars that point reached by a straight line extended outward from the center of the lunar globe through their footsteps. Whatever stars stand apart from that point will be considered as declining from the zenith of an observer stationed on the moon. This, therefore, is the basis for imagining an equator midway between the poles, and the declination of the sun from the zenith of places. Because the sun does not pass through the zenith of points on the equator every day of the year but only on the day of the equinox, the axis on which the lunar globe turns is not parallel to the axis of the ecliptic but inclines toward it; always, that is, the axis stands at right angles to the plane of the moon's orbit, which is inclined to the plane of the ecliptic at the same time that the moon's axis is inclined to the plane of the axis of the ecliptic.

and winter, but this cannot be compared with ours in variety,
nor does it always occur in the same places at the same time
of year, as is the case with us. For it happens that in the
space of ten years their summer moves from one part of the
sidereal year [time required by the earth to make one complete
revolution about the sun] to the opposite part, in the same
location; for in the cycle of nineteen sidereal years, or 235
Levanian days, toward the poles summer occurs twenty times
and winter the same number, whereas they occur forty times
at the equator;[101] they have annually six "days" of summer,
the rest winter, corresponding to our months of summer and
winter.[102] That alternation is scarcely noticed near the equator
because there the sun does not deviate more than 5° to
either side of the zenith at noon. It is felt more near the
poles, since the regions there either have sunlight or lack it
in alternate half-year periods, just as is the case on earth with
those who live at either of the poles. Thus the globe of
Levania also consists of five zones, which correspond in a

. . . . . .

[101] The nodes of the moon revolve in nineteen years with motion
retrograde to that of the sun. Therefore, the boundaries revolve in
the same time, as do also the poles of the moon's orbit (which are
the cosmic poles for the lunar people) in a small circle which
has a diameter of 5°. Therefore, the moon has twenty tropical
years in nineteen sidereal years. Thus, in nine and a half sidereal
years (ten tropical years) the tenth summer comes with the
sun in the constellation of Capricorn for those to whom in the
beginning of the period summer came while the sun was in the
constellation of Cancer. The same sort of thing happens to us on
earth, only much more slowly. Two thousand years ago, our summer came when the sun was in the constellation of Cancer and
the Dog Star rose with the sun; today our summer has passed into
the constellation of Gemini, although the zodiacal sign retains the
old name of Cancer.

[102] Of these six days only one or two are really summer; the
others on either side shorten toward the length of the equinoctial
day.

certain manner to our terrestrial zones; but the tropical zone
extends barely 10°, as do likewise the arctic zones; all the
rest are similar in extent to our temperate zones.[103] The tropical zone passes through the mid-parts of the hemispheres, half
its longitude, that is, through Subvolva, and the other half
through Privolva.

In the sections of the circle of the equator and the zodiac
there exist also four chief points, as we have the equinoctial
points [where the equator and ecliptic intersect—the ecliptic
being the central line of the zodiac, the great circle of the
celestial sphere that is the apparent annual path of the sun]
and the solstitial points [where the sun reaches its most north-
erly and southerly distances], and from these sections the
circle of the zodiac begins.[104] But the motion of the fixed
stars in natural succession from this beginning is very swift;
since in twenty tropical years (years, that is, designated by
one summer and one winter each) they travel through the
whole zodiac, which happens with us but once in 26,000
years.[105]

. . . . . .

[103] I can hardly call even these zones temperate. On the moon,
there is really no temperateness, as will be shown.

[104] The moon-dwellers have the zodiac in common with us of
earth. Our zodiac is defined by the annual movement of the earth
around the sun. The moon goes around our earth, even as we live
all around the earth. Therefore, we both have the same reason for
picturing the zodiac.

[105] This is related to what was said in note 101. It is appropriate
that concern for a tropical year, albeit not as great as our own
concern, be among the moon-dwellers at least this great [as indicated in the *Dream*] because of their tropical year. Consider
here the thesis of this work. Those features which are for us the
most important in the whole universe—the twelve heavenly signs,
the solstices, the equinoxes, the tropical years, the sidereal years,
the equator, the colures, the tropics, the polar regions, the cosmic
poles—all are confined to the very limited globe of the earth and

So much for the first motion.

The calculation of the second motions is no less diverse in the case of motions which they see than it is in the case of motions that we see, and it is much more complicated. For all six planets—Saturn, Jupiter, Mars, Sun, Venus, Mercury—experience, in addition to all the irregularities that are familiar to us, three others, two in longitude—one daily, the other in a cycle of eight and a half years—and the third in latitude, in a cycle of nineteen years. For in the mid-regions of Privolva the sun is larger when it is their midday than when it is their sunrise, other things being equal, and in Subvolva it is smaller;[106] the dwellers in both areas think that the sun deviates several minutes from the ecliptic in each direction, now toward these and now toward those fixed stars.[107] And these deviations have a pattern that is repeated, as I have said, in nineteen years. However, this wandering concerns Privolva

. . . . . .

owe their existence solely to the fancy of earth-dwellers. If we transfer to another globe, we must conceive of everything as changed.

[106] Because the orbit of the moon, that is, its distance from the earth at apogee, is one fifty-ninth of the distance between the sun and the earth. When, therefore, the Privolvans have the sun on their meridian they come closer to the sun than the earth does by one fifty-ninth of the whole; but when the Subvolvans have the sun they become farther away from the sun than the earth is. In fact, the full moon is distant from the sun by sixty parts, the earth by fifty-nine parts, the new moon by fifty-eight. As the interval diminishes, the sun appears larger. When the moon is in its quarters, it and the earth are equidistant from the sun. We have said that the sun rises and sets, when in quadrature, both in the midsections of Privolva and Subvolva.

[107] Because the ratio of the orbits is a little greater than 60:1, the moon when viewed from earth departs from the ecliptic about 5° on either side; but when seen from the sun, the moon's departure from the ecliptic is about 5'.

slightly more and Subvolva slightly less.[108] And although by
the first motion the sun and fixed stars are considered as proceeding with even gait around Levania, nevertheless, for Privolva the sun at midday advances hardly at all in relation to
the fixed stars, but to those in Subvolva it is very swift at
noon; the contrary is true at midnight. And thus the sun seems
to make certain leaps under the fixed stars, separate leaps on
separate days.[109]

. . . . . .

[108] When the dwellers in Privolva see the sun on their meridian,
they are closer to the sun than the earth is; when the inhabitants
of Subvolva see the sun on their meridian, they are farther away
from the sun than the earth is by a little more then a sixtieth part;
therefore, the Subvolvan-dwellers are farther away than the Privolvans by about a thirtieth of the whole. If the sun declines for
the Privolvans by at most 5′ 30″ it will decline for the Subvolvans
by 5′ 20″. I do not mention these things because I consider this
variation large or remarkable, since in truth it is almost impossible
for us on earth to see a sixth part of a minute; I mention them
in order to remove the suspicion that there might be from this
motion some greater variation of the latitude of the moon. If I
retained the same ratio of orbits that Ptolemy, along with the
ancients, gives, the declination would exceed about 15′.

[109] Because the earth and the moon proceed around the sun in an
annual motion, while the moon goes also around the earth, the
moon comes between the sun and the earth when it appears new
to us and at that time goes opposite to the motion of the earth.
It does not, however, go opposite to earth as much as earth
progresses in the other direction. For the earth daily moves
through a 365th part of its orbit, the moon through only a 30th
part of its orbit. The latter [the orbit of the moon] is a little
more than a 60th part of the former [the orbit of the earth]. A
30th part of the 60th part is about an 1800th part of the whole
terrestrial journey, and a fifth of 1/1800 is 1/365. Therefore, when
the moon appears full to us it is traveling 6/5ths as far as the earth
is traveling; and when the moon is new, 4/5ths; thus the latter
motion is two-thirds of the former. But the reader is warned that

These same deviations occur in the motions of Venus, Mercury, and Mars; in the case of Jupiter and Saturn they are almost imperceptible.[110]

Yet this diurnal motion is not consistent at identical hours each day, but sometimes it is slower, both in the case of the

. . . . . .

this *Dream* was written before the final perfection of the ratio of the orbits, when I was still agreeing with the ancient authorities that the sun is about 1,200 earth radii distant from the earth, the moon about 60 radii. Thus the ratio of the orbits is not 60:1 but 20:1. Since, then, the orbit of the moon is considered to be a 20th part of the orbit of the earth, the 30th part of this 20th (that is, the moon's daily part) is 1/600th of the earth's orbit and so more than half the daily motion of the earth. The daily progress of the earth under the fixed stars, then, leaves to the new moon a remainder of less than half, which is almost nothing; but to the full moon an excess of more than one and a half; that is, the full moon progresses more than four times as fast as the new. However, since the moon itself is thought of by the moon-dwellers as motionless, the sun instead will be thought of as moving at this uneven speed. This passage is cited in note 152 below.

[110] This daily motion of the earth and moon in a great circle around the sun, resulting from the annual motion of the earth and the monthly motion of the moon, is quite simply transferred to the sun (itself really motionless) by earth- and moon-dwellers. It is not so simple to transfer motion to Mercury, Venus, or Mars, for these have movements which seem peculiar to each. If the earth and the moon were to cease their annual movement completely, and the moon were also to cease its monthly motion, the planets nevertheless would seem to move, Mars through the whole zodiac very slowly when with the sun and very swiftly when opposite the sun, and Venus and Mercury not through the whole zodiac but in the vicinity of the sun would seem with alternating motion now to precede and now to follow the sun by a fixed number of degrees. These motions by which the earth and the moon are carried along seem to earth- and moon-dwellers to be connected with phenomena peculiar to the planets.

sun and of the fixed stars, and then faster in the opposite part of the year in exactly the same hour of the day.[111] And that slowness, making a complete circuit in the space of a little less than nine years,[112] makes its way through the days of the year, so that now it takes place on a summer's day and now on a winter's day, which in another year had experienced swiftness. Thus, now the day, now in turn the night, is made longer by a natural slowness, and not, as with us on earth, by the unequal division of the circle of the natural day.[113]

If the slowness befalls the Privolvan region in the middle of the night, an excess of night over day is accumulated, but if it happens in the day then night and day become more equal, which comes about once in nine years; the situation is reversed in Subvolva.[114]

. . . . . .

[111] If, as a result of the deceptiveness of phenomena of vision, the motions of the moon are transferred to the stars, certainly among the motions of the moon is one which makes it slow at apogee, swift at perigee. This happens in every successive phase: now the full moon is very slow, now the new moon, now the half moon. But when the moon is full, those who live in the mid-region of Subvolva consider the hour noon; when the moon is new they consider it midnight; it is the opposite for those in Privolva.

[112] So great is the motion through the zodiac at apogee.

[113] It being assumed that a straight line drawn through the center of the earth and meeting at right angles a line connecting the centers of the sun and the earth would divide the moon's orbit into arcs, the Privolvan day from the Privolvan night, for example, certainly the greatest inequality of the moon at quadrature is $7\frac{1}{2}°$, which when doubled is 15°. Therefore, the arc which has apogee in its middle is completed in an interval of 195°, the remainder in an interval of 165°. It is as if you were to say that an earth night is thirteen hours long, an earth day eleven hours long, or the reverse. But they have a different number of hours.

[114] This slowness comes from the position of the moon at apogee.

So much then for those features which are in a certain manner common to the two hemispheres.

CONCERNING THE PRIVOLVAN HEMISPHERE

Now as to the features that are unique in each hemisphere, there is great difference between them. It is not just that the presence and absence of Volva result in such different spectacles, but even the common phenomena themselves have very different effects in one region from what they have in the other, to such an extent that perhaps it would be more correct to call the Privolvan hemisphere nontemperate and the Subvolvan hemisphere temperate. For in Privolva the night is as long as fifteen or sixteen of our natural days; it is gloomy with perpetual darkness, like that of our moonless nights, for it is never illuminated by any rays of Volva; therefore, everything is stiff with cold and frost,[115] and there are besides very strong, sharp winds;[116] there follows a day as long as fourteen

. . . . . .

The dwellers in the midregions of Privolva have their midnight at the time when we earth-dwellers see the full moon. If, therefore, the full moon and apogee coincide, the night is lengthened for the Privolvans; but if the moon is new at apogee, the days are more nearly equal to the nights, the opposing causes nullifying each other.

[115] If you assume that there are living beings on the moon, you will concede that for sustaining and fostering them there are evaporations from the body of the moon; but the vapor, thin and surrounded by cold, is congealed into snowy powder, which is the formation of frost.

[116] In a dream one must be allowed the liberty of imagining occasionally that which never existed in the world of sense perception. Thus one must assume here that winds exist because globes stir the ethereal air—a cause which I remember I did not reject when I was discussing the reasons why morning is more pleasing and salubrious to all things that live and breathe on the

of our days, or a little less,[117] when the sun is quite big[118] and slow-moving with respect to the fixed stars[119] and there are no winds.[120] Accordingly, there is immeasurable heat. And thus for a space of one of our months, which is a Levanian day, there is in one and the same place heat fifteen times more burning than our African heat and cold more intolerable than of Quivira.

It must be especially noted that the planet Mars, in the mid-regions of Privolva, in the middle of the night, and in

· · · · · ·

earth, and also why there is everlasting snow on the highest summits of many mountains even in the torrid zone.

[117] I attribute to the Privolvan day only fourteen of our days, and to their night fifteen, because the lines AE and AD drawn from the center of the sun A to points E and D of the lunar orbit separate the exterior part of it (EGD) from the interior (ECD) and establish the former as longer than the latter by about 4 degrees in accord with the measure of the ratio of the orbits. In the whole exterior arc the midregions of Privolva are in the shadow of the moon. At the two points E and D they are lit by the first and last rays of the sun. And they receive the sun's rays throughout the whole interior arc of orbit ECD (see fig. 1, p. 56). [Note: See reproduction of Kepler's fig. 1, page 134 of this work.]

[118] From the earth we see the sun in a size of 30′. The moon, in the new phase, is nearer to the sun than we on the earth are by about a 59th part or a little less. Therefore, in that hemisphere of the moon which is then illuminated by the sun, the sun appears slightly bigger, that is, by about half a minute. The ancients thought that the ratio of the orbits was much smaller, in fact a ratio of 1:18, which would amount to a little less than 2′.

[119] As in note 109 above. In fact, the sun is reckoned as being a third slower at the meridian in the midregion of Privolva than it is at the meridian of the midregion of Subvolva.

[120] Assuming those things that were assumed in note 116. For certainly the new moon is more gently rubbed against the upper air than is the earth, by a fifth, and more gently than the moon itself is when the moon is full, by a third.

other Privolvan areas, each in its own part of the night, appears almost twice as big as it does to us.[121]

CONCERNING THE SUBVOLVAN HEMISPHERE

Passing over to this hemisphere, I shall begin with those who dwell on its border, that is at the divider circle. For it is peculiar to them that they see the elongations [angular distance] of Venus and Mercury from the sun as being much greater than we do.[122] At certain times, even Venus appears

. . . . . .

[121] Almost twice, I say. The apogeal distance of the sun from the earth is 101,800. The perigeal distance of Mars from the sun is 138,243. Therefore, if the earth at aphelion and Mars at perihelion should be joined in the same longitude, there would remain an interval of 36,443 from earth to Mars. Assume now, in accord with the opinion of the ancients, that the diameter of the moon's orbit is exactly an 18th part of the diameter of the sun's orbit, and assume also that the moon is full so that Mars will be closest to the Privolvan region when it is midnight in Privolva. An 18th part of 101,800 is 5,655, the distance by which the Privolvans are closer to Mars than we earth-dwellers are. The ratio of this number (5,655) to 36,443 is less than a sixth. Hence the Subvolvans at the time of our new moon, when their Volva is full, would see Mars less than a third smaller than would the Privolvans at the time of our full moon, which is their midnight.

In the truer ratio of the orbits which I used in my *Rudolphine Tables*, this proportion is reduced. The distance from the moon to Mars is not as much as a 21st part of the distance to the sun. Thus the difference in the appearance of Mars to Privolvans and Subvolvans will be a little less than an 11th part of the whole.

[122] When the moon's distance from the sun is not much different from the earth's distance from the sun, the digressions of Venus and Mercury from the sun can also be observed by those who dwell in the midregion of Subvolva. But for those who live on the divider the sun appears on the horizon when the moon is either full and farthest from the sun or when it is new and closest

twice as big to them as it does to us,[123] especially to those
who live near the northern pole.[124]

. . . . .

to the sun. The digressions of the planets, especially the digressions
of Mercury, are perceived either just before sunrise or just after
sundown. In the true proportion of the orbits, the difference in the
rectilinear intervals between the sun and the moon is about a
30th of the whole, wherefore the difference in the intervals of the
digressions, too, is not much different.

[123] In order that Venus may appear larger to moon-dwellers
than to earth-dwellers, it is necessary for Venus to be nearest the
earth and for the moon to be nearest the sun. But when the moon
is nearest the sun at the time of our new moon, those who live
in the midregion of Subvolva see neither the sun nor Venus because the hour is midnight. Therefore, this appearance is left to
those who inhabit the divider. Moreover, the variation in the
appearance of Venus is a little more evident to the dwellers of
Subvolva than the variation of Mars is to the Privolvans (although the inhabitants of the divider can see both), because in
the nearest approach of Venus to the earth there remains an interval between them of 25,300, of which (being less than the
36,443 interval between earth and Mars) the diameter of the
moon's orbit is the larger portion.

[124] The divider circle was defined above as going through the
poles of the moon's monthly revolution. Now the orbit of the
moon has latitude in one direction in the north, in another direction in the south; and the axis, whose extremities are the poles, is
considered to touch the plane of the eccentric orbit at right angles.
Therefore, although neither of the poles of the moon inclines more
toward the sun than the other, it happens nevertheless that the
pole of our ecliptic, which is regarded by them as the mean ecliptic,
is different from the pole of the moon's orbit; indeed, the lunar
pole goes around the earth's pole once in the space of nineteen
years. Hence, when a position of Venus between earth and the
sun is required, Venus must be seen not through elongation of
longitude, but only through latitude. Its southern limit is in the
sign of Pisces, its interval at aphelion not much ahead in the

But the most pleasant thing of all in Levania is contemplation of Volva, the sight of which the dwellers there enjoy in
place of our moon, which is entirely lacking to those in both
Subvolva and Privolva.[125] And the Subvolvan region is so
designated from the constant presence of Volva just as the
other is called Privolva from the absence of Volva, because
the dwellers there are deprived of the sight of Volva.

When our moon is rising full and advancing above the
distant houses, it seems to us earth-dwellers to equal the
circumference of a wine cask; when it ascends to the meridian,
it displays the breadth scarcely of a human countenance. But
when Volva is in the middle of the sky (a position it has for
those who live in the very center, or navel, of this hemisphere), it seems to the dwellers in Subvolva a little less than
four times greater in diameter than our moon does to us, so
that if one were to compare the two discs, Volva would be
fifteen times bigger than our moon.[126] But to those for whom

. . . . . .

beginning of Aquarius; nevertheless, it is then seen from the earth
and from the moon in the opposite signs of Leo and Virgo. If,
then, the pole of the moon inclines toward these signs of the
mean ecliptic, it inclines toward the north and so makes Venus
more clearly and accurately visible through its latitude, for the
latitude is greater when it [Venus] is at aphelion than when it is at
perihelion.

[125] That is, they lack a view of the moon pursuing its course
among the stars. Since they inhabit it (as we are now imagining)
the moon-dwellers see the moon as we see our earth.

[126] This is a matter of the apparent diameters, not the true ones.
The apparent radius of the moon at apogee is 15′, but the parallax
of it in the same position is 58′ 22″, which is a little less than 60,
the quadruple of 15. As great as is the parallax of the moon, so
great would the radius of the earth appear to an eye in the moon.
Therefore, the ratio is a little less than 4:1, which when squared
is a little less than 16:1, that is, greater than 15:1 for the apparent
discs. Thus:

Volva hangs perpetually at the horizon, it presents the appearance of a mountain on fire at a distance.

Therefore, just as we distinguish regions according to greater or lesser elevations of the pole (granted that we do not perceive the pole itself with our eyes), so the altitude of the ever visible Volva, differing from place to place, serves the same purpose for them.

For, as I have said, Volva stands overhead at the zenith for some, and for others appears low near the horizon; for the rest, it varies in altitude from the zenith to the horizon, but always, in any given place, the altitude is constant.[127]

They also have their own poles, however;[128] these are not

· · · · · ·

58′ 22″ Logistic logarithm	2761	(Sixtieth part of *Chilias*.
15′ 0″ Logistic logarithm	138629	Vol. VII, pp. 391 ff.)
The ratio is	135868	
Double it so that it is	271736	

This number, as the logistic logarithm shows, is 3′ 58″. Hence, if the disc of the earth is 60′, that of the moon is 3′ 58″; since 4′ 0″ is a fifteenth of 60, the ratio is therefore a little greater.

[127] Because the moon always turns the same spots toward the earth, a line connecting the centers of the earth and the moon always intersects the surface of the moon in the same spot. And the inhabitants of that spot have our earth (that is, their Volva) always at the zenith. However many degrees of a great circle any place is removed from this spot, by that same number of degrees in the sky does Volva seem to decline from the zenith.

[128] There is a certain rotation of the lunar body in the space of a month. In its whole circuit it turns the same face to the earth, as is proved to us by the perpetually unchanging spots. However, since the earth (that is, Volva) seems to go in the space of a month through the whole zodiac, so also does the face of the moon go with it, turning now toward Cancer and now toward the opposite sign Capricorn. That is, the moon rotates. To those who are on the moon, however, the moon does not seem to turn but seems to be at rest, just as to us the earth seems to be still. In-

at those fixed stars which mark our cosmic poles in the sky,[129] but are in the region of other stars which are for us indicators of the poles of the ecliptic. In the space of nineteen lunar years, these poles of the moon-dwellers move in small circles around the poles of the ecliptic,[130] in the constellation of Draco and the opposite constellations of the Swordfish (Dorado) and of the Sparrow (Flying Fish) and of the greater

. . . . . .

stead of the moon rotating, the sky seems to turn in the opposite direction. It follows, then, that there are also two points of the sky around which, as if immovable, the sky seems to them to turn once a month. These points are called poles.

[129] If the axis of the moon remained parallel to the axis of the earth throughout the whole circuit, we would sometimes see new spots around the northern and southern edges of the moon, that is, at the time when we see the moon opposite the sun in Cancer or Capricorn. For a line drawn from the center of the earth through the boundary of the torrid zone, and meeting the zodiac at either of the solstitial points, intersects the axis of the earth at unequal angles; it would also intersect an axis of the moon which was parallel to the earth's axis at the same angles; wherefore, each of the poles of the moon would be exposed to our sight at opposite times of the year. Since this does not happen, the axis of the globe of the moon is not parallel to the axis of the earth but is always intersected at right angles by a line from the center of the earth. The moon's axis does not, therefore, tend toward the same points toward which the axis of the earth tends. The axis of the earth tends toward what we call the cosmic poles; the axis of the moon does not tend toward those poles.

[130] The poles of the revolution and monthly circuit of the moon are not the same as the poles of the ecliptic. The former go around the latter in little circles which have a radius of 5°, and they complete one circuit in reverse order in the space of nineteen years. Since, therefore, these lunar poles do not depart more than 5° from the poles of the ecliptic, they are rightly said to be around them and thus, also around the fixed stars which indicate the poles of the ecliptic.

nebula. Since these poles of the moon-dwellers are distant
from Volva by almost a quarter of a circle, their regions can
be marked off both according to the poles and according to
Volva.[131] It is obvious how they outstrip us in respect of convenience; for they indicate the longitude of places by means
of their unmoving Volva,[132] and the latitude by means both
of Volva and the poles,[133] while we have nothing for longi-

. . . . . .

[131] The moon in its motion describes the orbit with whose poles
we are concerned. Yet the poles of every great circle on a sphere
are always distant from all parts of the circumference by exact
quadrants. Hence the orbit of the moon is not a perfect circle.
For in quadrature the latitudes of its orbit are enlarged; that is,
the orbit extends farther toward the poles at its syzygies and thus
is not then distant from those poles by quite a whole quadrant.
This actual curving path of the moon around the earth, so to
speak immovable in the center of the universe, is transferred by
the inhabitants of the moon (who fancy their own home at rest)
to the earth, their Volva.

[132] It was said in note 127 that every great circle which goes
through Volva finds it (Volva) unmoving, and that the different
degrees of the circle are distinguished by the varying altitude of
Volva. Among the great circles, however, is one which proceeds
midway between the poles of the moon from west to east. Wherefore even in that a distinction of place can be made according to
the degrees of altitude of Volva toward west or east. But this is a
difference of meridians, or of longitude.

[133] Indeed, the altitude of the pole of the lunar regions provides
the same aid to observation as on earth, or not much different.
This altitude of the pole can serve equally for the latitude of
places under all meridians because all lunar meridians meet at the
poles of the moon-dwellers. The altitude of Volva can be very
readily observed, yet it does not serve equally at all meridians for
establishing the latitude of a place. Only on that central meridian
of all, which goes constantly through Volva, and only in the semi-
circle of it which divides the midregion of Subvolva, does the
altitude of Volva immediately disclose also the latitude of a place.

tudes except that insignificant and scarcely discernible declination of our compass needles.[134]

For them, then, Volva stands in the sky as if fixed by a nail, and above it the other stars and the sun pass from east to west.[135] There is no night in which some of the fixed stars in the zodiac do not betake themselves behind Volva and emerge again on the opposite side.[136] However, the same fixed stars

. . . . . .

Outside this true primary meridian, there is need to adduce proof in addition to the altitude of Volva—its elongation from the primary meridian. This is the same reasoning we earth-dwellers use when we seek the altitude of our pole from the altitude of the sun at equinox but outside the meridian.

[134] We do indeed have eclipses and occultations of the fixed stars by the moon, but that is a very laborious and hazardous method. When I was writing *Lunar Astronomy*, the declination of the magnet from the meridian was in some repute as being universally suitable for proving the latitudes of places. About that time, the *Mecometria* of a certain Frenchman had been published. But William Gilbert's theory of magnetism and his repeated experiments, when very carefully examined, proved the Frenchman's efforts insignificant and empty. There is no fixed point on the globe of the earth, except the pole, to which the magnetic needle points; but the needle is somewhat attracted to high mountains everywhere.

[135] You have here the full statement of my primary thesis. We earth-dwellers consider that the level surface on which we stand, and along with it the turrets on our church towers, rest motionless while the stars move past those turrets from east to west. But this idea restricts the truth not at all nor does it bring an exception against it. Likewise, the inhabitants of the moon think that their lunar landscape and the globe of Volva suspended above it are still, whereas we know with certainty that the moon is a moving body.

[136] This is true not only of very small, inconspicuous stars but even of the brightest ones in the first rank. For to them one night

do not do this every night;[137] all those which are within 6° or 7° of the ecliptic[138] take turns. A cycle is completed in nineteen years, at the end of which time the first stars return to the place they had in the beginning.[139]

Volva waxes and wanes no less than our moon,[140] the cause being the same: the presence of the sun or its departure. Even the time is the same, if you look to nature; but they reckon in one way, we in another. They consider as a day and a night

. . . . . .

is as long as fifteen of our nights, and during its passage a 25th part of the zodiac, that is 14°, is seen to pass behind Volva, because in truth, there are about twenty-five half-months in one year. But in the space of 14°, some suitable fixed stars certainly appear, inasmuch as the sky everywhere is sprinkled with them.

[137] Whatever the moon cuts off from us earth-dwellers at any time, the same is cut off from the moon-dwellers by their Volva, that is, our earth, at an opposite point of time. Look at my *Ephemerides*, which I published for future years, and you will see at the end this alternation of the fixed stars (Vol. VII, pp. 618 ff). For us, any one fixed star is occulted by the moon once during most months of the year; during the following year that star remains free and another star is occulted in its turn.

[138] The center of the moon departs from the path of the ecliptic by 5° 18′ at most. To this arc about 15′ must be added on account of the parallax, because the radius of the moon is a little greater than one-fourth of the radius of the earth. Now the apparent radius of Volva is almost four times the apparent radius of the moon; consequently, it is as much greater as the parallax is smaller, in the same ratio as the true radii. Thus on both sides is gathered a total of about 6⅓°.

[139] The same fixed stars actually return more often but not in the same order until after the cycle is completed. The cause of this alternation is the circuit of the nodes of the moon in reverse order in the same number of years.

[140] See the diagram suitable for observing this, *Epitome of Copernican Astronomy*, page 560 [Vol. VI, p. 364 of *Opera Omnia*].

the time in which all the waxings and wanings of their Volva
occur. We call this amount of time a month. Because of its
size and brightness, Volva is almost never, even in its new
phase, concealed from the Subvolvans.[141] It is especially visible
in the polar regions, which lack the sun at the time. There,
at midday, Volva turns its horns upward.[142] In general, for

. . . . . .

[141] Frequently with us the moon, when new, is visible for scarcely
a few hours after it has passed beyond the rays of the sun—not
only in the illuminated crescent but throughout the whole body.
It has been established that this phenomenom results from the
earth's illumination of the lunar hemisphere which is turned away
from the sun, for the earth then turns its whole orb, illuminated
by the sun, toward the moon, and reflects the light of the sun
onto the moon. Surely analogy teaches that the moon-dwellers will
see no dissimilar things in the case of their Volva, which is our
earth. Assume that it is the time of new Volva; it will be assumed
in turn that the moon is full for us earth-dwellers. Everyone knows
with what splendor the full moon paints the earth (which is the
Volva of the moon-dwellers), especially on winter nights when the
moon proceeding in Cancer illuminates our land from on high.
It is not, therefore, nonsense to say that the surface of the earth
(which is the Volva of the lunar-dwellers) is made visible as far
as the moon by that very lunar light which brings the mountains
and neighboring plains of earth out of the night into our view.
For although the moon receives and reflects scarcely a 15th part of
the light which the earth receives from the sun, the face of the
earth itself is fifteen times bigger in the sight of the inhabitants of
the moon than the moon is to us earth-dwellers. Therefore, there
is a compensation. Add now the bright crescent of Volva, which,
on account of sideward digression from the path of the sun, appears fifteen times more clearly to the moon-dwellers than the
crescent of the moon appears to us earth-dwellers. It is to this
crescent of Volva that the words "almost never" have a special
reference. When there is no breadth there is no crescent left at
the moment of conjunction.

[142] In other places on the moon, the sun and Volva appear together at the moment of new Volva. If, because of the brightness

those living between Volva and the poles on the mid-Volvan circle, new Volva is the sign of midday, the first quarter the sign of evening, full Volva of midnight, the last quarter of sunrise.[143] For those who have both Volva and the poles situated on the horizon, and live where the equator and the divider intersect, morning or evening comes with new and full Volva, noon and midnight with the quarters. From these

. . . . . .

of the sun, the thin crescent of the moon which remains because of latitude is not perceived on the earth, that is due to the nature of vision. For those who live between the poles of the path of the moon and the path of the sun, the sun at midday is always below the horizon, the moon a little above the horizon, and therefore so much the more clearly visible.

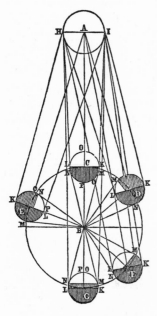

FIG. 1. Kepler's figure 1, as reproduced in Frisch's *Opera Omnia*.

[143] Consider that day is beginning. The sun is standing on the horizon. The moon is in the first quarter. When the angle AEB, connecting the centers of the sun A and the earth B with the center of the moon E, is a right angle, the angles of AL touching the body of the moon will be almost right angles. Now, if a ray from the center of the sun touches the surface of the earth, the sun will be regarded as on the horizon of that place of earth where the ray strikes. For the same reason, the sun will be considered on the horizon also at the point on the moon L, which is touched by AL, and a line drawn from the center E of the moon through point L to the earth will mark the zenith of that place on the moon. Therefore, those who have Volva at the zenith have the sun on the horizon; that is, they have the beginning of day.

examples one may make judgments also concerning those who
live in between.[144]

They even distinguish the hours of the day according to
this or that phase of Volva: the closer the sun and Volva
come together, the closer noon approaches to the former and
evening or sunset to the latter. During the night, which regularly lasts as long as fourteen of our days and nights, they are
much better equipped to measure time than we are. For apart
from that succession of the phases of Volva of which we have
said full Volva is an indication of midnight to the region of
mediavolva, Volva itself also marks off the hours for them.
For although it appears not to move at all from its place,[145]
unlike our moon, it nevertheless turns like a wheel in its
place[146] and displays a remarkable variety of spots one after

. . . . . .

[144] On the other hand, in the region of those moon-dwellers to
whom Volva appears on the horizon at the time of the quarter,
as in O, the rays of Volva BO touch the globe of the moon, and
a straight line EO drawn from the center of the moon E out
through point O makes a right angle at O with BO; therefore,
since this point spreads out into almost a complete great circle
on the lunar body there will be some point on the circle through
which a straight line drawn from E will meet the ecliptic. Let that
point be O. But for places lying under the ecliptic, the poles of
the ecliptic are on the horizon. If it is the time of the quarter,
when the angle EOB is a right angle, it is necessary that at such
a place as O, which has the poles of the ecliptic and Volva on
the horizon, line EO should come very close to the sun and thus
the sun should hang at the zenith and the hour should be noon.

[145] The globe of the earth, traveling through the zodiac in the
space of a year, does move in space. But it seems to the moon-
dwellers to remain fixed in space because they have no aids for
sensory perception of this motion. They think that it is rather
the sun which, in their twelve alternations of day and night, makes
the revolutions in an opposite direction. The same thing happens
also to us earth-dwellers with reference to that same sun.

[146] For the globe of the earth also turns once around its axis in

the other, these spots moving along constantly from east to west.[147] One such revolution, in which the same spots re-

. . . . . .

the space of a day. This motion of the earth is displayed before the eyes of the moon-dwellers, and they have no reason to believe that the globe of Volva is not turning on its axis or to think that the whole sky (as commonly believed among us), and with it their home, the moon, goes around Volva, contemplating one after another of the regions of the Volvan globe, although the latter is in fact the truth. Let us take the appearance of sunspots as an example. We see them going around the sun's body in about the space of twenty-six days. Who would conclude that the sunspots are in fact motionless, but that the boat which we call earth carries us around the sun in such a short space of time and reveals the various parts and spots of the sun's surface to us one after another? Followers of Copernicus are convinced that they are carried around the sun in the space of a year. For this reason they are certain that they could not make the trip in twenty-six days, for this is a contradiction. Sight affords us the surest proof of the revolution of the sun. Visual evidence also proves to the moon-dwellers that their Volva turns on its axis. Let their visual perception deceive, or let it be confirmed—the result is the same whichever you choose; it is firmly established that the lunar-dwellers, if there are any such, must be convinced of the revolution of Volva. Q.E.D. So far as it pertains to the more secret goal of this allegory, a pleasant retort occurs to us. Everyone says it is plain that the stars go around the earth while the earth remains still. I say that it is plain to the eyes of the lunar people that our earth, which is their Volva, goes around while their moon is still. If it be said that the lunatic perceptions of my moon-dwellers are deceived, I retort with equal justice that the terrestrial senses of the earth-dwellers are devoid of reason.

[147] From the spots on the moon we arrive at conclusions about the formation of the surface of the moon—that it is a combination of water and dry land. Nor are these idle conclusions. For we demonstrate by means of very certain optical principles that roughness and smoothness of surface are connected with variations of spots and brightness; those parts which are bright are high and

turn,[148] is considered by the Subvolvans as one hour of time,[149] although it equals somewhat more than one day and one night of ours.[150] And this is the only constant measure

· · · · · ·

hilly, those which are dark are flat and low-lying. A distinction between land and water follows. Such are the thoughts that we earth-dwellers have about the surface of the lunar globe.

By a reversal of this same process of logic, I allow my moon-dwellers to see earth's mountains and seas as spots of brightness and darkness (see note 154).

[148] The surface of the earth turns on its axis with respect to its center from west to east; but from the point of view of spectators dwelling on the moon, the surface of the earth that is turned toward the moon seems to go from east to west in accordance with that axiom in Aristotle's *Mechanics* which says that opposite parts of a circle (or globe) seem to go in opposite directions when viewed from outside the circumference of the globe.

[149] If the presence of the sun is to be called day and its absence night, the rather long period of the sun's stay above the horizon of places on the moon certainly requires a subdivision into smaller parts. For if our day with its night, which is only the 29th part or a little less of a lunar day and night, is subdivided for the sake of convenience into twenty-four parts, how much more need is there for dividing the much longer period of the lunar day. Nature has deserted us earth-dwellers in that we are not able to discern with our eyes those matters which the mind and judgment of man behold; there is nothing anywhere which, revolving, returns to its original position in the space of one of our hours. But on the moon the dwellers in Subvolva have before their eyes the motion of Volva around its axis, which brings back the spots of Volva in the same order fourteen times each night. It is not at all probable that this observation is neglected by the lunar people. From it, we can estimate in some measure the poverty and solitude of the Privolvan people who, deprived of the sight of Volva, lack also this aid in marking off time.

[150] We must here distinguish three different measurements of completion of a revolution of the surface of earth, or Volva. The first is the time required for a given point on earth's surface to re-

of time.[151] For, as was said above, the daily journey of the sun and the stars around the moon-dwellers is irregular, as is shown perhaps especially by this rotation of Volva, if it were to be compared with the elongations of the fixed stars from the moon.[152]

In general, as far as its upper, northern region, is concerned, Volva seems to have two halves.[153] One is darker and covered

. . . . . .

turn to a position under the same fixed star; this is a little less than one of our natural day-and-nights. The second measurement is the time required for a point on the surface to return to a line through the centers of the sun and the earth; this clearly equals and also causes a natural day-and-night. The third measurement is the time required for a point on the surface to return to a line from the center of the earth through the center of the moon; this is the revolution that finally brings the spots of Volva back into view on the moon. Compute the length of this period as follows: in seventy-six years there are 940 lunations while the fixed stars are completing a circuit of 1,465 nineteen times; that is, 27,835 revolutions, or 10,020,600 degrees. In this space of time, 27,759 of our days elapse. So take away 940 lunations; 26,819 days remain. That many times do the same spots on Volva return to the view of the moon-dwellers. Divide the total of the degrees by 26,819; the result is 373⅔. That is the number of times the equator slips by while the moon-dwellers see one complete procession of the spots on Volva. Thus one hour on the moon is as long as a little more than one and one-thirtieth of our natural day-and-nights— that is, almost twenty-five of our hours.

[151] Indeed, this same revolution of the earth engenders in us earth-dwellers the fancy of a *primum mobile* whose motion is conceived of by astronomers as always uniform.

[152] See note 109 for both the fact and the cause of this unevenness of the sun's motion, as the moon-dwellers observe it.

[153] The two halves are the old world (with Europe, Asia, and Africa joined together) and the new (North and South America). I have restricted the differentiation of the halves to their northern portions because the region of Magellan, very extensive in the

with almost continuous spots,[154] the other is a little
. . . . . .
south, is unknown and thought of as being a continuous continent
reaching to both hemispheres, both of the New World and of the
Old World.

[154] I mentioned this passage in the *Conversation with the Star
Messenger* of *Galileo,* which I published at Prague in the year
1610. At the same time I corrected it because Galileo taught me
that the high and rough parts of the moon are not spots but
brightness, and that water flowing into lower areas becomes dark
and takes on the appearance of spots. The same judgment, therefore, must be made with regard to the earthly globe: that the
ocean and seas surrounding the lands take on darkness, that the
continents and islands shine brightly in the light of the sun. I
had at first come to the opposite opinion because the surface of
the earth took on various colors while water was thought of as
lacking color. But every color (except white) is a step toward
blackness. Indeed, the reflected brightness of sunlight is in proportion to the darkness of the surfaces whence it is reflected.
Water supplied another argument. Whenever one looks at surfaces of land and water placed next to each other, the land is
always dark and the water shines. See the experiment in *Optics*
made while I was viewing the river Mura from a certain Mount
Schockel in Styria (p. 251). I gave as the cause of the brightness
the mirrorlike smoothness of the surface of the water, the roughness of the land. I was much occupied with consideration of these
causes in chap. 1 of *Optics.*

At present, I must refute those arguments (which I did briefly
in *Conversation,* p. 15) and I must establish, by means of arguments, Galileo's contrary opinion which I quoted in *Conversation.*
So far as the colors of land are concerned, you would be more
correct, or at least equally correct, in saying that all colors except
black are steps toward pure light. As for the lack of color in
water, Aristotle denies this in his work *On Colors,* where he ex-
pressly maintains that the color of water verges on black. He
uses an argument from a visual phenomenon—all land appears
darker when it is wet with rain, but when the moisture is dried
out by the warmth of the sun the land gleams more brightly. I

. . . . . .

added another practical experiment when at Prague a certain learned man was standing near me on a bridge and pointing out to me the gleam of the water in order that he might tear Galileo's argument to shreds. I bade him look at the reflections of the houses in the water and compare their appearance with the appearance of the real houses, for the difference in brightness is obvious—the reflections in the water are darker. In such a manner was my former argument about the colors of the earth and of water dissolved and turned around. As to the second argument from the reflected brightness, it is of such a nature that I destroyed its force in the same *Optics* in another place where I concerned myself with the illumination of the moon. If we apply the example of water viewed from near at hand to round bodies at great distances, we stray very far from our path and allege as a cause that which is not a cause. For water flowing near land shines not with its own brightness but from the brightness of the air which is illuminated by the sun, whose bright rays falling on all sides are reflected to our eyes. Let there be put above the water a sail, which cuts off the brightness of the lower air. Straight away, the gleam of the water is extinguished. I put this dissolution of my argument in the margin of p. 252 of *Optical Part of Astronomy* when I was rereading it. Heavenly bodies which are illuminated by the light of the sun and viewed from a great distance are by no means seen by reason of rays reflected according to the laws of optics and of mirrors but by a light actually transmitted by the sun, as I said in *Optics*, and appropriated to the bodies by reason of the roughness of their surfaces. This communicated light is by virtue of its definition stronger on land than on water. This suffices as a refutation of the contrary argument. As for the true opinion, which holds that the spotted parts are such things as seas and lakes, the bright parts dry continents or islands, you have the evidence for this plainly set forth in Galileo's *Star Messenger* and in my *Conversation* with him, p. 16, and finally *Copernican Astronomy*, Vol. VI, p. 831, and above note 147.

Having first said these things as a necessary caution, I shall now set forth the reason for the individual parts of this description. First of all I made the appearance of the ancient world darker, brighter,[155] with a shining girdle flowing between and dividing the two halves to the north. The shape is difficult to explain.[156] In the eastern part[157] is what looks like the front of a human head cut off at the shoulders,[158] approaching a young girl [159] with a long dress,[160] to kiss her. She, with hand extended backward,[161] is enticing a cat that is jumping up.[162] However, the bigger and wider part of the spot[163] extends westward without any clear shape.[164] In the other half of Volva, the brightness is more widespread [165] than the spot.[166]

. . . . . .

thinking, as I said, that land is dark. I said that the spots are almost continuous because Europe is joined with Asia in Scythia, Asia with Africa in that part of Arabia which is between Egypt and Palestine.

[155] I said that the appearance of the half containing the New World was a little brighter (as a result of the error mentioned above) because it has more seas and large areas of ocean, internal as well as external, which squeeze America at the middle into a narrow isthmus and, so to speak, strangle it.

[156] The Brazilian, the Atlantic, the Deucaledonian, the Arctic oceans, extending to the Anian Strait and going to the Japanese ocean and the ocean of the Philippines, the Moluccans and the Solomons.

[157] With respect to the so-called belt, or the Atlantic ocean.

[158] Africa.

[159] Europe.

[160] Sarmatia, Thrace, the Black Sea region, Muscovy, Tartary.

[161] Britain.

[162] Scandinavia, or Denmark, Norway, Sweden.

[163] Asia, Tartary, Cathay, China, India, etc.

[164] Of course Asia extends eastward from Europe. But because the moon takes the same way around the earth as the surface of the earth takes around its axis, the lower hemisphere of the earth or Volva seems to the lunar-dwellers to go to the west.

[165] Both oceans (according to the mistaken hyopthesis, of course).

[166] The American continent.

You might say that there was the likeness of a bell [167] hanging
from a rope[168] and swinging westward.[169] What there is
above[170] and below[171] cannot be likened to anything.[172]

Not only does Volva in this way mark off the hours of the
day for them, but it even gives clear evidence of the seasons
of the year to anyone who is paying attention and who fails
to notice the positions of the fixed stars. For even at the time
when the sun occupies Cancer,[173] Volva clearly shows the
northern pole of its revolution. Appearing in the middle of
the bright area,[174] above the figure of the girl, there is a
certain small dark spot,[175] which moves from the highest and
outermost part of Volva[176] toward the east, and thence, descending into the disc, toward the west,[177] from which extreme

. . . . . .

[167] South America.

[168] Nicaragua, Yucatan, Popayan.

[169] See note 164. For of course Brazil faces east toward Africa.

[170] North America.

[171] Brazil.

[172] The region of Magellan.

[173] When the sun is in Cancer, the pole of the earth or of the
primum mobile (that is, of the revolution of Volva) withdraws
by only $66\frac{1}{2}°$ from the center of the sun and so also from the
center of its own disc, which the moon-dwellers see along a
line through the center of the sun and of Volva. The disc of Volva,
therefore, extends beyond its pole by $23\frac{1}{2}°$, seen at an angle.
Whatsoever has a disc with a radius of 60′ has a line extending
from the center of the disc to a point under the pole.

[174] In the Northern Ocean.

[175] Thule, or Iceland. (But according to the false hypothesis, that
the dry parts of the earth's surface are darker than the wet.)

[176] The outermost edge of the disc of Volva touches the polar
circle. Iceland lies near the circle. Wherefore, in any one revolution
of Volva it goes once to the outermost edge of the disc when the
sun is in Cancer.

[177] As in notes 169 and 164.

it moves again back to the upper part of Volva toward the east and so is then always visible.[178] But when the sun is occupying Capricorn, this spot is nowhere visible, the whole circle, with its pole, being hidden behind the body of Volva. And in these two seasons of the year the spots head in a straight line for the west,[179] but in the times between when the sun is in the east (that is, established in Libra) the spots descend or ascend transversely in a somewhat curved line. From this evidence we know that while the center of the body of Volva remains still, the poles of this rotation travel in a polar circle around their own pole once in a year.[180]

. . . . . .

[178] The converse of this is valid. If the sun in Cancer is visible constantly to all the inhabitants of the polar circle throughout the whole revolution of the earth, the arctic circle always will be visible likewise either to the sun or to an eye situated, as the moon-dwellers are, on the moon in a line extending to the centers of the sun and of the earth.

[179] A plane which goes through the centers of the moon and the earth and intersects the circumference of the ecliptic at right angles also crosses then through the poles of revolution of Volva. But when the sun is at the equinoctial points the pole of Volva stands to the side of this plane. The equator of Volva is then intersected by it at oblique angles. It is amusing that this same thing is observed also in the spots of the sun, as I wrote in a letter to Bartsch in the year 1629 concerning observations made by the most illustrious Prince and Lord, Duke Philip, Landgrave of Hesse.

[180] The lunar-dwellers, not knowing that they are carried along together with Volva in an annual motion under the fixed stars, must ascribe annual motion to the poles of Volva. For although the axis of the earth tends toward the same fixed stars through the whole year, the poles of earth are distant from the pole of the ecliptic. And this distance is maintained unaltered as the earth travels through the ecliptic along with the moon. But sometimes the earth approaches those fixed stars which mark the pole of Volva, and sometimes it departs from them. Consequently, the

The more diligent notice also that Volva is not always the same size. For in those hours of the day in which the stars are swift, the diameter of Volva is much greater, so that then it is more than four times as large as that of our moon.[181]

Now what shall I say about eclipses of the sun and of Volva, which occur also on Levania and at the same moments in which eclipses of the sun and the moon occur here on the globe of the earth, but plainly for opposite reasons? For when we see the whole sun go into eclipse, Volva goes into eclipse for them; when, in turn, our moon goes into eclipse, the sun is eclipsed for them.[182] Not everything matches exactly, however. For they often see partial eclipses of the sun when we are having no eclipse of the moon,[183] and, on the other hand, they often experience no eclipse of Volva when we are having partial eclipses of the sun.[184] Eclipses of Volva occur for them

· · · · · ·

position of the pole of Volva seems to move in turn from this side to that side of the pole of the ecliptic and thus to revolve around it.

[181] The variation in the diameter of Volva as seen from the moon is just the same as the variation we inhabitants of earth ascribe to the parallax of the moon. Therefore, the radius of Volva at apogee is 58'22", at perigee (when the sun is swift) 63'41", since for us the radius of the moon at apogee is 15', four times which is 60'.

[182] The moon obscures the sun for us and we darken the moon (that is, the globe of our earth does). Similarly, the moon-dwellers (or rather their moon) shadow our earth, their Volva; here their Volva, our earth, takes the sun away from them.

[183] Nevertheless, the moon then is pale for us also, especially in the part nearer the shadow.

[184] When the center of the penumbra, which is generally occupied by the true shadow of the moon, does not enter the disc of the earth, or when it does enter but there is no shadow of the moon—only a residual circle of the sun. Although in the first case the moon-dwellers see no full shadow in the disc of Volva, they do see a certain lightness and darkness in the extremities,

at the time of full Volva, as eclipses of the moon do for us at
the time of the full moon. But eclipses of the sun take place
at the time of new Volva as for us at the time of the new
moon.[185] Since they have such long days and nights, they
experience very frequent darkenings of both heavenly bodies.
Whereas with us a great part of eclipses cross over to our
antipodes, their antipodes, which are in fact Privolva, see
absolutely nothing of these; the dwellers in Subvolva alone
see them all.

They do not ever see a total eclipse of Volva,[186] but they
see a certain small spot,[187] which is ruddy at the edges,[188]

. . . . . .

where the penumbra falls. If, in the second case, the indicated
center of the shadow does pass through, they see a partial shadow
around the center as if it were cast by a thin cloud or transparent
veil within indistinct boundaries, similar to shadows cast by tower
turrets on the plains below—not solid, but interspersed with sunlight.

[185] But do not forget that at the moment when we have a new
moon they have a full Volva. When we have a full moon they
have a new Volva.

[186] For the disc of the earth (that is, Volva) has a radius between
63' 41" and 58' 22", while the shadow of the moon, which is the
cause of eclipses of Volva seen by the moon-dwellers, diminishes
on account of the size of the sun to a radius never greater than
1' 22" by the time the shadow reaches the earth from the moon.
And often the shadow is nonexistent at that distance.

[187] Never greater, that is to say, than a 46th part of the diameter
of Volva.

[188] On account of the diminution of the solar rays. I had reference to what happens in a closed room when the sun shines
through a very narrow aperture. But there the border usually shows
red because complete shadow surrounds it and the light of the
sun is within the border and the comparison is more obvious. In
the disc of Volva, however, the shadow of the moon is within and
is of the most insignificant size while the whole disc of Volva
glows outside. Thus the redness, in comparison with the surround-

and black in the middle,[189] go across the body of Volva; this spot enters from the east of Volva and leaves by the western[190] edge, taking the same course in fact as the spots that are native to Volva, but surpassing them in speed,[191] and it

. . . . . .

ing brightness, is necessarily faint. See how anxiously I concern myself with making corrections lest some recent witness of these matters should come down from the moon and prove me wrong.

[189] Everyone must know, from the experience of looking down from high places on a summer's noon, what is the appearance of a shadow on a level piece of ground illuminated by sunlight. What we earth-dwellers thus behold at close range is the same as what moon-dwellers view under the name of Volva from a distance. However, on account of the quality of air (that is, upper air) around the sun, the shaded place will be darker when absolute night descends on us during an eclipse of the sun, which happens occasionally. See *Epitome of Copernican Astronomy*, p. 895.

[190] As in notes 164, 169, 177, above. For the surface of the earth, and the moon standing above it facing the sun, and the shadow of the moon falling on the surface of the earth (which we are imagining that the lunar-dwellers observe), all move in one and the same direction.

[191] See the diagrams of solar eclipses, considered as a whole, at the beginning of my *Ephemerides*. They were prepared for the

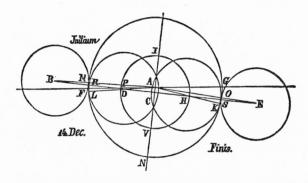

FIG. 2. Kepler's figure 2, as reproduced in Frisch's
Opera Omnia.

lasts a sixth part of their hour or four of our hours.[192]

Volva is the cause of their solar eclipses, as our moon is the cause of ours; since Volva has a diameter four times larger than that of the sun, it is a necessary consequence that when the sun moves from the east, through the south behind the motionless Volva, and to the west, it very frequently goes behind Volva and thus the sun's body, either in part or as a whole, is hidden by Volva. The occultation of the sun's body, moreover, frequent though it is, is nevertheless very remarkable because it lasts several of our hours,[193] and the light of

. . . . . .

purpose of representing these eclipses of the moon-dwellers' Volva; still, in those diagrams also an eye is imagined in the moon as necessitated by the demonstrations. In the period of an hour, 15° of the earth's equator (in the middle of the disc of earth A) will revolve. But half of 1° of the moon's globe will go across, and the shadow of the moon will darken a little more of the disc of the earth, for example, PC. But one-half of the lunar globe equals almost 60 halves on earth, as a result of the ratio of the diameters of the globe of the earth and the orb of the moon. Therefore, during one of our hours the shadow of the moon PC passes over more than 30° of the terrestrial equator on the earth's disc while the surface of the earth's globe itself is rotating through only 15°. Thus the shadow PC moves twice as fast as the parts of the earth nearest the center of the disc A which are facing the moon-dwellers. But the shadow in R or S is incomparably swifter than in the downward sloping parts of the earth's equator in F and G, the extremities of the visible disc.

[192] The *Rudolphine Tables* had not yet been worked out at that time. However, see examples of almost central conjunctions in the year 1633 on April 8 and October 3; the times correspond exactly. But remember that the farther from the center the shadow of the moon crosses, the greater is its stay on the disc.

[193] An eclipse of our moon is an eclipse of the sun for the lunar-dwellers. But an eclipse of the moon, from beginning to end, can last 4 hours and 20 minutes, and the entire moon remains in the

both the sun and Volva is extinguished at the same time. It must certainly be an impressive thing for the dwellers in Subvolva, whose nights otherwise are not much darker than days on account of the splendor and size of the ever present Volva, when, during an eclipse of the sun, both luminaries, the sun and Volva, are extinguished for them.

Their eclipses of the sun have this singular feature which happens quite often; scarcely is the sun hidden behind the body of Volva when from the opposite side appears a shining, as if the sun has become enlarged and embraced the whole body of Volva, when otherwise the sun regularly appears so much smaller than Volva.[194] There is, therefore, not always complete darkness, unless the centers of the bodies are almost completely aligned,[195] and the arrangement of the transparent

. . . . . .

shadow of the earth 2 hours and 8 minutes. See *Epitome of Copernican Astronomy*, p. 868. Therefore, the entire sun can likewise be hidden from the lunar-dwellers for the same number of hours.

[194] You have in my *Optics*, in the chapter about lunar eclipses, a diagram in which I show the refraction of the rays of the sun in the atmosphere surrounding the earth, where the refracted rays enter the boundaries of the shadow on the eastern side, for example, continue through the depths of the shadowy cone, and leave by the western side. Hence the moon, approaching the western boundary of the shadow, meets the refracted rays of the sun coming from the eastern edge of the earth. These become visual rays and the moon-dwellers think they see a small eastern part of the sun beyond Volva even when almost the whole sun is visible in front of Volva on its eastern side. And this happens in those places of the moon which we see as particularly reddish during an eclipse of the moon, for this redness is caused by the refracted rays of the sun.

[195] And unless the moon is proceeding through its apogee. For observational experiments have demonstrated that the refracted rays of the sun cross the shadow lower down and do not overtake the moon at apogee.

middle part [that is, atmosphere] allows it.[196] But Volva is not
so suddenly extinguished that it cannot be seen at all,[197]
though the whole sun hides behind it, except just at the midpoint of a total eclipse.[198] At the beginning of a total eclipse,
at certain places on the divider, Volva still shows white, as if
a living coal remained after a fire has been put out; when this
whiteness is also extinguished it is the middle of a total
eclipse (for in a partial eclipse the whiteness remains), and
when the whiteness of Volva returns (at opposite points of
the divider circle), there comes a view also of the sun; so
that thus in a certain manner both luminaries are extinguished at the same time in the middle of a total eclipse.[199]

. . . . . .

[196] There is without doubt sometimes present in the vaporous
substance a certain marvelous light which does not come from
the sun, either through its primary rays or its secondary rays. This
happens in our terrestrial atmosphere so it can also happen in the
lunar air.

[197] For the earth, that is, Volva, is illuminated also by the full
moon and receives from this illumination a certain whiteness.
Therefore, as long as only one or other edge of the moon, the
eastern or western part of the divider circle (and not the whole
moon) is deprived of the sun's light, Volva, along with the sun,
does not become entirely invisible. The whole sun is indeed hidden by Volva from that particular edge of the moon but Volva,
by reason of the white light it receives from the moon, is in
turn visible to the dwellers on that edge of the moon. This is confused writing and should not be taken to refer to an eclipsing of
Volva but to the ordinary extinguishing of Volva at the time of
new Volva, like the extinguishing of our moon at the time of
every new moon.

[198] Do not take this "middle" to apply to any particular place on
the moon, but to the whole period when the moon tarries in the
shadow of the earth. For then the moon, lacking light, sheds no
light on earth (that is, Volva) which then is cutting off the whole
sun from the whole moon.

[199] Take the words thus also: Let major and minor eclipses for

So much for the phenomena in both hemispheres of Levania: Subvolva and Privolva. It is not difficult to judge from this, even without my saying anything, how much the Subvolvan hemisphere differs from the Privolvan hemisphere in other respects.

For although the Subvolvan night is as long as fourteen of our day-night periods, nevertheless the presence of Volva brightens the land and protects it from cold. Certainly such a mass and such brightness cannot fail to produce heat.[200]

On the other hand, though, the Subvolvan region has the troublesome presence of the sun for fifteen or sixteen of our

. . . . . .

the moon-dwellers correspond to total and partial eclipses of the moon for earth-dwellers. In all these, apply the word eclipse to the sun, but to Volva only to the extent that, on account of the failing sun, sunlight reflected by the globe of the moon onto the earth (that is, Volva) also fails. Thus Volva lacks the primary light of the sun because of the usual new Volva period, and lacks also the secondary light of the moon because the sun is eclipsed for the lunar-dwellers.

[200] We can put the heat of the lunar light (although this is scarcely a 1/15th part of the light of Volva) to proof by the sense of touch, helped, however, by art. For if you catch the rays of the full moon with a concave parabolic mirror, or even with a spherical one, you will feel at the point of focus, where the rays come together, a certain warm exhalation as it were. This happened to me at Linz when I was concentrating on other experiments with mirrors and was not thinking about the heat of light. I began to look around to see if anyone was breathing on my hand.

There is no need of proof that this light of Volva (that is, of our earth illuminated by the sun) is in the category of heat producers, since sometimes in the summer there is such intense heat from the rays of the sun that forests and wooden buildings catch fire, whereupon the crowd makes accusations of arson. What of it, then, if the moon is fifty thousand miles away from this heat, since the distance permits a simultaneous view of nearly a whole hemisphere of the terrestrial globe.

days and nights; still the sun, being smaller, is not dangerously strong,[201] and the luminaries in combination lure all the water to that hemisphere,[202] and the land is submerged, so that very little stands up above the water,[203] while the Privolvan hemisphere, on the other hand, is dry and cold, since, of course, all

. . . . . .

[201] This in iself is very little, but it should perhaps not be neglected in the amassing of causes. For the sun is farther away by a whole diameter of the lunar orbit from the Subvolvan region at the time of new Volva than it is from the Privolvan region in the middle of the Privolvan day.

[202] It is a probable conjecture, but there is no full demonstration. Experienced seamen say that the ocean's swell is greater in the syzygies of the heavenly bodies than in the quadratures. But the causes of the ocean's swell seem to be the bodies of the sun and moon which attract the waters of the sea by a certain force similar to magnetism. The body of the earth also attracts its own waters unto itself, by a force which we call gravity. Why, therefore, should we not say that the earth also attracts the waters of the moon as the moon attracts terrestrial waters? This being granted, if now the sun and Volva come together or are opposite each other, their attractive power will be combined. Since, however, the joined bodies cling for a long time at the zenith of the Subvolvan regions and do not depart as quickly as they do from the zeniths of the terrestrial oceans, there seemed to be enough time for drawing all the water from one hemisphere of the moon to the other. Something is lacking from this reasoning, however. In order that this may happen, it is necessary for the whole surface of the moon to be accessible to the waves and that no shore be anywhere barring the way. But the telescope has shown us mountains, hills, and vast seashores. Those ramparts must be broken through with valleys and channels like deep gorges in order that such a quantity of water may be able to go back and forth from one hemisphere to another. Let us believe this for the time being, until some explorer goes into the matter at first hand.

[203] There is no likelihood that the mountain tops, pressed so close together at such an altitude, will be submerged.

the water has been drawn away.[204] When, however, night approaches the Subvolvan area, and day comes to Privolva, since
the luminaries are divided between the hemispheres, so also is
the water, and the fields of Subvolva are laid bare, but moisture is provided to Privolva as a slight compensation for the
heat.[205]

All Levania does not exceed fourteen hundred German
miles in circumference, only a fourth part of our earth.[206]
Nevertheless, it has very high mountains,[207] and very deep and

· · · · · ·

[204] Then, moreover, the Privolvans have midnight and the Subvolvans have Volva (although it is not conspicuous) along with
the sun. If we compare this with nautical observations made on
earth, we find the opposite: that, in fact, in the middle of the
night, when the luminaries are absent, the flux of the ocean is
just as great as it is at midday when the luminaries are present.
Prophecy falters here, therefore, unless you would ascribe the nocturnal flux of our oceans to a rebound from the shores of America,
against which the moon, drawing the waves after her, dashes them,
and in turn from the shores of Europe and Africa, in reciprocal
fluctuation, which the moon, returning the next day, guides with
a new command. You must remove from the moon such beating
of the water against the shores, which is the cause of the fluctuation of the waves, if you wish to deprive the Privolvans of all water
in the middle of their night.

[205] Of course, it is said that when the moon is in quadrature, the
flow and ebb of the sea is almost imperceptible, as if there came
about an equalization of attractive power between the sun rising
and the moon departing from the middle of the sky, or the other
way around.

[206] The diameter of the earth, and so also a great circle of the
earth, are to a diameter and a great circle of the moon as 389 is to
100 (*Epitome of Copernican Astronomy*, p. 483). The diameter
of the moon is therefore a little greater than a fourth part of the
diameter of the earth.

[207] This small section of the *Dream* is older than the Dutch telescope. I hold Maestlin, my teacher in astronomy, entirely responsi-

broad valleys,[208] and thus falls far short of earth in perfect

. . . . . .

ble for it. It is a part of the theses which I mentioned above in note 2. And I repeated it also in my *Optics*, p. 250. Use of the telescope confirms it to a remarkable degree, as do certain of Galileo's observations which I mentioned in my *Conversation* (p. 20) and certain observations of my own: projections of land rising perpendicularly from the surface of the earth five thousand paces begin to be visible at a distance of forty-five German miles (see *Epitome of Copernican Astronomy*, p. 23). But if you should travel all over the navigable oceans, you would hardly find a longer distance from which land is seen; therefore, no mountain rises above the surface of the water more than a great German mile. See Snell's *Dutch Eratosthenes*. As to the number of mountains on the moon, and their height, and as to the interval, in the half-moon, by which points from the bright section extend into the dark part (reaching from the shadowy depths to the light of the sun's rays) see Galileo's *Assayer*, written against Sarsi, and other works by Galileo. In the year 1612, in the month of May, while I was observing an eclipse of the sun, a ray coming through a telescope with double lens fell on a white tablet. I saw on the circumference of the moon's shadow (that is, the eclipse which the interposition of the moon was causing in the shape of the sun), I saw, I say, in this concave circumference, two very clear protuberances extending beyond the roundness of the shadow, that is, of the moon, into the bright concave figure. Lest you say that these protuberances were phenomena of the lens or some trick of vision, I point out that they stayed on the disc of the sun and traveled through it at the same speed at which the moon was moving, and they preceded the shadow of the moon in leaving. If you ask me the ratio of their altitude to the moon's diameter, I say that it would have been impossible to perceive them had they not been equal to at least a 60th part of the moon's diameter. For the ray I was observing was very small, not much bigger than a silver coin of the Empire. The mountains on the moon were, therefore, at least eight miles high, because the diameter of the moon is about five hundred miles.

[208] Below, in the appendix, you will find [mention of] a pit, per-

roundness. Moreover, the whole of it, especially in the Privolvan tracts,[209] is porous and pierced through, as it were, with
hollows and continuous caves[210] which are the chief protection of the inhabitants against heat and cold.[211]

. . . . . .

fectly round, as if it had been made by hand, ten German miles
in diameter. In the middle of it there is a huge winding fissure,
such as I think the valley of the Enns crawling through Mount
Caecius or the valley of the Inn through the Alps would appear if
one were to view them from high up when the sun is sinking. But
this pit on the moon is comparatively much deeper and craggier.
So that you may not lack cause for amazement, it seems to be
crossed at one point by a less shadowy passage, almost a sort of
bridge. These observations are more recent than this book. So
much the more delightful is this anticipation of the truth,
"manly beyond its years in mind and countenance."

[209] Which no eye has ever seen. Nevertheless, you see the reasoning in the mention of Privolva, where the logic is especially valid
because of the great intemperateness of climate and the most violent alternation of extreme heat and cold.

[210] This was not a simple prediction, resulting from consideration
of the terrific heat of such a long day, in order that I might make
the moon habitable even for living lunar beings. I had made con-
jectures about the loose texture of the moon's body from its motion, and I analysed these in my *Commentaries on Mars*. The following year Galileo's *Message from the Stars* appeared; with very
clear observations, in which he likened the moon with its network
of caves to the tail of a peacock, he rendered this theory more
valid. See my *Conversation* with this *Messenger*, p. 14. This is the
subject of the whole letter which I wrote, with proofs, as an appendix to this *Dream*.

[211] Right here is the reason, leaving aside visual proofs. If I had
then regarded it as an established fact that the moon has as many
low-lying areas as Galileo's telescope has brought to light, or if I
had read Plutarch's tale of the cave of Hecate, I believe that I
should have asserted my theories with a freer pen.

Whatever is born from the soil or walks on the soil [212] is of prodigious size. Growth is very quick; everything is short-lived, although it grows to such enormous bodily bulk.[213] The Privolvans have no settled dwellings, no fixed habitation; they wander in hordes over the whole globe in the space of one of their days, some on foot, whereby they far outstrip our camels, some by means of wings, some in boats pursue the fleeing waters, or if a pause of a good many days is necessary, then they creep into caves; each acts according to its nature. Most creatures can dive; all breathing beings naturally draw their breath very slowly; and, therefore aiding nature by means of art, they live deep down under the water.[214] For they say

. . . . . .

[212] Lunar land, I mean. I was figuring that living creatures are in proportion to the mountains. See the *Optical Part of Astronomy*, p. 250. The comparison with our earthly creatures is not one of bodies only, but also of functions: respiration, hunger, thirst, waking, sleeping, working, resting. Evidence of this can be seen in the continual excesses of heat and cold, the rare recovery of life, and the large size of the works mentioned in the appendix. Concerning this, see Plutarch's work, p. 1730.

[213] This is from the Tübingen theses. It, too, is according to the proportional analogy, to which I have been very attentive since an early age. For us here on the earth, the very slow motion of the fixed stars, the brief orbital periods of the individual planets, and the daily revolution of the earth itself seem to me to be related to the length of the human life and to the moderate size of human bodies. Since for the moon the fixed stars return more quickly than Saturn, and since the lunar day on the other hand is thirty times longer than ours, I thought I should ascribe to the lunar creatures a short life and huge growth so that nothing would come to a standstill but everything would perish in the midst of growth. In the theses I passed over into politics, believing that public affairs are subject to repeated and very great changes but that private fortunes are frequently great.

[214] Since I had taken the water away from Privolva, and was

that in those deep recesses of water, cold water remains, while the upper waves are heated by the sun,[215] and whatever clings to the surface is boiled by the sun at midday and becomes food for the approaching swarms of wandering inhabitants.[216] For, on the whole, the Subvolvan hemisphere is comparable to our villages, towns, and gardens; the Privolvan hemisphere is like our fields and woods and deserts. Those to whom breathing is more necessary introduce the hot water into the

. . . . . .

forced to leave it enormous alternations of heat and cold in swift succession, it occurred to me that those regions would not be habitable, at least not under the open sky. It was very convenient, then, to have the water flow in at fixed points of the day and I made the living creatures accompany the water as it receded again. That they might do this the more speedily, I gave long legs to some and to others I gave the ability to swim and the capacity to survive immersion in water but only to the point that they did not degenerate into fish. None of this will be incredible to anyone who has read about Cola, the Sicilian man-fish. I was thinking that there is no extreme on earth so violent but what God gave a tolerance of it to a particular species of animals: tolerance of hunger and African heat to the lions, tolerance of thirst and the vast desert of Palmyrene in Syria to the camels, tolerance of hyperborean cold to the bears, etc.

[215] On this basis: that all matter, insofar as it lacks life, is of itself cold; and if it grows warm from an outside cause, it becomes cool again of its own accord when the alien heat departs as the cause of the heat is removed. Now in Privolva the surface waters certainly feel the heat of the sun, but the rays of the sun do not, I believe, penetrate to the deepest recesses of the water because of the depth.

[216] Everything certainly must have been intended for a definite use. The heat of the water followed from their very long day. Such effects have been observed in the province of Chile under the circle of Capricorn and in the torrid zone even with our brief earthly day; for they write that the rain comes down distinctly warm.

caves by means of a narrow canal, in order that, being taken
into the innermost parts through a long course, the water may
gradually become cool. They stay there for the greater part
of the day, and use the water for drinking; and when evening
approaches they go forth to seek food.[217] In the case of plants,
bark, in the case of animals, skin, or whatever may take its
place, accounts for the major part of the corporeal mass, and
it is spongy and porous; if anything is caught in the daylight,
it becomes hard and burnt on top, and when evening approaches the outer covering comes off.[218] Things growing
from the soil, although on the mountain ridges there are few,
are usually produced and destroyed on the same day, new
growth springing up daily.

A race of serpents predominates in general. It is wonderful
how they expose themselves to the sun at midday as if for
pleasure, but only just inside the mouths of caves, in order
that there might be a safe and convenient retreat.[219]

Creatures whose breath has been exhausted and life extinguished through the heat of the day return at night, in a
contrary fashion to the way flies do with us.[220] For scattered

. . . . . .

[217] This is, as it were, their occupation. If anyone should object
that those regions are uninhabitable if even the water is hot, I'll
send him down into our cellars and deep wells where we cool our
beverages in summer.

[218] According to the diverse provision of nature. My precedents
here were the husks and rinds of our plants and fruits, the shield-
shaped shells of oysters and turtles, the hard parts of the feet of
living creatures—the hooves and soles.

[219] I had read in Arnobius the African that it was among the
pleasures of his peoples to expose themselves naked to the sun and
to bask like lizards, and if I am not mistaken also like crocodiles.
Since these last are native animals, I thought their habits might
be partaken of by the Africans. But this would be rather in the
nature of a torture for us Europeans.

[220] Thus it is written about the people of Culomoria, a province

here and there on the ground are masses in the shape of pine nuts, which have parched shells in the daytime, but in the evening, when the hiding places open up, as it were, they put forth living creatures.[221]

The chief alleviation of the heat in the Subvolvan hemisphere is the constant cloudiness and rains,[222] which sometimes prevail throughout half the region or more.[223]

. . . . . .

of Northern Scythia: that there are those who die when the long night is upon them but are restored to life when the sun returns. Because of this, they seek a safe entombment, lest any harm befall their bodies during the absence of their souls. See what Martin Del Rio says about these people in *Investigations of Magic*.

[221] Scaliger, in *Exercises*, says that resin sweated from the timber of ships in the heat of the sun forms drops which give birth to ducks whose bills develop last of all the body. When the bills are free, the ducks give themselves to the waves below. There is in Scotland a tree of famous report that puts forth similar offspring. In the year 1615, in a very dry summer, I saw at Linz the branch of a juniper tree which had been brought from the abandoned fields of Traun. Attached to the branch was growing a strangely shaped insect of the color of a horned beetle. The beetle had risen as far as its middle from the branch and was moving very slowly; the hind parts clinging to the tree were juniper resin.

[222] Jose d'Acosta writes the same about the provinces of the New World. See my *Conversation with the Star Messenger*, p. 18.

[223] I derived this conjecture from a disputation which Maestlin defended, and which was published in the year 1605 under the title *The Phenomena of the Planets*. I deal with this also in *Conversation*, p. 19. Because of its relevance, however, it would be a good thing to review the whole matter here. Therefore, let us digress to it for a little while. In his theses 136 and 143, the author begins to deal with the following occurrence: that the moon sometimes appears twice on the same day, old in the morning and new in the evening, when it cannot be more than 6° or 7° distant from the sun; although on other occasions 12° are required for its reappearance. In thesis 146, he proposes a new theory to explain

. . . . . .

this phenomenon; that the moon is enveloped by a certain airy substance. For in thesis 139 he had proved that when it is 12° distant from the sun the moon has scarcely 1/80th part of its visible diameter illuminated by the sun. How much less of the moon will be illumined, then, if it is not more than 7° distant from the sun? He concludes that the whole of this atmosphere which extends beyond the boundary of the body of the moon is colored by the rays of the sun, since it is permeable by them, that is, transparent. Thus the moon is never completely extinguished, even in the very central conjunction. He confirms the theory with five additional abservations. First, when a ray from the sun undergoing eclipse has passed through an aperture, it always makes the outer convex circumference in the sun's image bigger than the inner concave circumference: that is, than the dark part cut off, or covered by the convex body of the moon. This happens even though the diameter of the full moon is generally gerater than the diameter of the sun. Consequently, he thinks that when we measure the full moon we include in the measurement that part which stands out all around the body and which consists of illuminated lunar atmosphere; but when the moon covers the sun we do not see this atmospheric garment because the rays of the sun pass through it unimpeded and unrestrained.

This observation, gained by watching a solar eclipse, is in fact very convincing. It caused Tycho Brahe also to say that the diameter of the new moon is smaller than the diameter of the full moon, and Longberg in his *Danish Astronomy* agrees with his teacher. David Fabricius, too, the Frisian astronomer whose opinions I discussed in the introduction to my *Ephemerides,* was much concerned with this nightshirt of the moon. It is, I say, very true that, in the image of the eclipsed sun which enters through a small aperture, the convex circumference is part of a bigger circle and the concave circumference is part of a smaller circle. But the reason alleged by the disputant is not the proper reason. I do not intend to deny the existence of a lunar air. I accepted it in my *Optics,* pp. 252, 302, and in *Conversation,* p. 18. It does not, however, accomplish what the disputant seeks. For the observation has another explanation: the radius of the aperture through which the

. . . . . .

ray of the sun enters. A shining border as wide as this radius is joined to the sickle-shaped sun all around, even in the points of the horns which thus become blunted. When stripped of this shining border, the pure image is left with the outer circumference shrunken and the inner concave circumference expanded. If this correction is applied, the diameter of the moon when it covers the sun is found to correspond to the diameter of the full moon.

The man who was maintaining the opposite side in the debate quotes this solution of mine from my *Optics*, which was in print at that time. In a note to the thesis, he mentions and accepts my denial of the difference between the new moon and the full moon. He does not, however, eliminate this first observation from a number of observations of lunar air. I suppose he thought that the reader should be permitted to judge for himself. Or does he think that he used the smallest possible opening, through which even the horns of the image were rendered sufficiently sharp? I certainly do not believe that. For there is a great difference between the ratio of the diameters which the author adduces from an observation of a solar eclipse 1605 2/12 October, and the ratio which I find in similar observations. Observers must be warned, therefore, that the paper on which the image of the eclipsed sun is being projected must be guarded against the slightest motion, and it must always be placed at right angles to the aperture, and it must be perfectly flat. For if the paper bends, the circumferences of the shining image are distorted and degenerate from circles into ellipses. Let the disputant determine whether he has been sufficiently careful to eliminate this difficulty.

The occurrence which has been adduced as the explanation for the lessened diameter, I do not deny. So it is necessary for me to explain why that occurrence cannot be the cause of the diminution. Without doubt the reason is that even transparent objects cast shadows when placed in the sun. I proved this in *Optics* by means of experiments with a glass ball full of water. The water transmits the rays of the sun and concentrates them to such a degree that they burn clothing and ignite powder. But the transmitted rays are deflected to another place, and the edges of the ball meanwhile are casting their own shadow in straight lines from

. . . . . .

the sun. If, now, no shadow falls from where the light of the sun can pass through, what will happen in lunar eclipses, which we often see, when both luminaries are above the horizon? The light of the sun passes through our air here and makes its way even to the moon, the earth being no hindrance at all because both luminaries are overhead. What, then, is this thing which envelops the moon in shadow at that time if it is not our atmosphere impeding the direct rays of the sun? So much for this first proof of lunar atmosphere.

The second observation of the air around the moon is in thesis 148. When the darkened part of the half-moon begins to occult some star, this dark part is seen nearer the center of the moon than the opposite shining limb. When the full moon is about to hide the stars, it is seen first to take them up in the embrace of its bright tunic and they shine through it; then it hides them behind its body and covers them completely. You have an observation of this type in the *Rudolphine Tables*, precept 133, page 94, concerning a conjunction of the moon and Venus. There is also a fourth observation of this same type in thesis 150: that when the whole body of the crescent moon is visible with a pale and weak light next to the shining sickle, then, I say, the circumference of the shining sickle appears wider than the opposite circumference of the body. The disputant thinks that the bright light of the sickle is due to the amplitude which extends beyond the body. Add also a fifth observation from thesis 151: that the crescent of the moon is estimated never to be narrower than one digit, although sometimes on the same day the moon is perceived both full and new, with the illuminated part occupying scarcely an 80th part of the diameter. The disputant again contends that this airy tunic is seen extending beyond the boundaries of the body.

I did not consider these three observations to be suitable evidence for such a wide extension beyond the body of the moon. I ascribed the cause of the phenomenon rather to the nature of vision. For at night the pupil of the eye naturally dilates. A shining point of light makes a stronger impression and imbues a wide area of the netlike membrane of the eye with visual impulses. The same thing happens by day when the eye is directed

. . . . . .

toward a strong light. In this way the image of the visible objects is distorted in the netlike membrane, since the bright parts expand into the adjacent dark regions. But this image on the concave retina within the eye is an exact inversion of the visible object on the outside. In a note to thesis 151, the author acknowledges this solution, too, without mentioning my name, but rejects it on the grounds that the same thing happens also in the daytime. But the phenomenon that I use to refute the argument, though more evident by night, nevertheless prevails also by day.

Nevertheless, some evidence of lunar air is presented in these theses, especially in the fourth and fifth observations. For because the rays of the sun pass through the lunar air and make it very bright, the limb of the moon, although otherwise by no means capable of casting a shadow, vigorously stimulates vision by means of the brightness which it has absorbed; this vigor permeates the retina and accounts for the appearance of so great a width of brightness in the visible object, thus causing the timely emergence of the moon. It is not, I say, the actual width of the visible sickle that is commensurate with the apparent width. But the strong effect of the real brilliance on the eye gives an illusion of excessive width. See my arguments along this line against the not dissimilar observations of David Fabricius in the introduction to my *Ephemerides*.

I have passed over the third proof, brought forward by the disputant in thesis 149. The margin of the shining moon is bright and pure and without spots, but the whole middle of the moon seems spotted. Certainly the cause is that the lunar air, thin in the middle of the body and shallow on the sides, is deep where it is seen at the edges. Similarly, on the level plains of earth, the atmosphere overhead (although illumined by the sun) does not stimulate vision very much; for those looking up from the bottom of a deep well, the air does not obscure the larger stars. Yet the air above distant mountain tops becomes white because the vision is penetrating to a great depth, and the more distant mountains are touched with blue or even obscured completely. When the sun is absent, the air obscures even the brightest stars as they rise. Thus, as a rule, there are either no clouds, or only a few thin ones,

When I had arrived at this point in my dreaming, a wind accompanied by the sound of rain came along and dissolved my sleep and destroyed along with it the last part of that book obtained at Frankfurt. And so, having taken leave of the Daemon who was speaking, and of the audience, Duracotus the son with his mother Fiolxhilde, even as they had their heads covered, so I came to my senses to find my head in fact covered by a pillow, my body wrapped in bedclothes.

. . . . . .

overhead, while toward the horizon they are always very thick, even when there are almost none overhead.

These are Maestlin's proofs of air on the moon; this is their force. After them, he states his thesis 152, the penultimate one of his book, in which he compares the lunar air with the atmosphere surrounding earth. He compares the brightness of the moon's limb, which has the same cause as the marvelous phenomena of our dawn. And he lifts our eyes on high, as I do to the moon, that they might recognize from there quite similar phenomena in the case of our earth.

Finally, he adds a note saying: "Whether that air, like ours, condenses into clouds, which by their opacity assume the appearance of very solid bodies, and as a result glow red and fiery, as with us when the sun rises or sets, I leave in doubt. This much experience has certainly taught us: that that surrounding brightness appears more or less clear at different times." And he adds evidence that is suited to my hypothesis: "In the year 1605, on the eve of Palm Sunday, in the body of the moon which was undergoing eclipse and which had assumed the color of red-hot iron, there was visible toward the north a blackish spot darker than the rest of the body. You would have said that over a wide region were spread clouds, heavy with showers and rainstorms of a kind we often see when we look down from mountains to places in the valley below." Sometime afterward, I had a talk with him, in the course of which he declared that that spot was of no common size but that it occupied more or less half the diameter. It was the recollection of this remark that brought to a close this last part of my *Dream*. With repetition of it, I likewise conclude my notes.

Geographical, or, if you prefer, Selenographical Appendix

To the Very Reverend Father Paul Guldin, Priest of the Society of Jesus, etc.
Venerable and Learned Man, Beloved Patron.

There is hardly anyone with whom I should prefer at this time to discuss matters of astronomy than with you, if, in addition to the pleasure of this conversation, there might be some other value to be gained from my trip at this troubled time when the whole court is beset with cares of war. Even more of a pleasure to me, therefore, was the greeting from Your Reverence which was delivered to me by members of your order who are here. Father Zucchi could not have entrusted his most remarkable gift—I speak of the telescope—to anyone whose effort in this connection pleases me more than yours. Since you are the first to tell me that this jewel is to become my property, I think you should receive from me the first literary fruit of the joy that I gained from trial of this gift.

Now what shall I say? If you will bring your mind to bear on towns on the moon, I shall prove to you that I see them.[1] Those lunar cavities which Galileo first noticed chiefly indicate spots;[2] these are, as I demonstrate, depressions in some level surface, as are our seas.[3] But I gather from the shape of the cavities[4] that these places are, rather, swamps.[5] In them, the descendants of Endymion make a practice of measuring out the spaces of their towns[6] for the sake of guarding them[7] against the mossy dampness[8] as well as against the heat of the sun,[9] perhaps even against enemies.[10] The method of fortification is this: They fix a stake in the center of the space to be fortified,[11] and to this stake they tie cords[12] which are long or short[13] according to the size of the future town; the longest I discovered is five German miles.[14] With this cord thus fixed they mark off the circumference of the future rampart[15] by the ends of the cords.[16] Then all gather together in full force to construct the rampart.[17] The moat is not less than one German mile in width.[18] In some towns they take all the excavated material [19] inside.[20] In others, they take it partly to the outside, partly to the inside;[21] thus there is a double rampart[22] with a very deep moat between.[23] The individual ramparts form a circle as round as if it had been drawn by a compass,[24] an effect achieved by the equal length of the cords extended from the central stake.[25] In this way it comes about not only that the moat is very deep, but also that the center of the town gapes in the form of a hollow, like the navel of a swollen belly;[26] and the whole circular border is raised high by the heaping up of the material excavated from the ditch.[27] For from the moat to the very center would be too great a distance to carry the material.[28] In this ditch, therefore, is gathered the moisture of the damp ground,[29] and whatever space is inside the ditch is drained by it;[30] and the moat, when overflowing with water, becomes navigable.[31] When the moat is dry, it is passable by a land road.[32] Therefore, wherever the heat of the sun assails those who are in the middle of the town[33] they betake themselves to that part of the circular moat which lies in the shadow of the outer rampart.[34] Those who are beyond the center seek the shade of the inner rampart[35] in the part of the trench away from the sun. Thus, during the whole fifteen days when the place is parched by the sun, they follow the shadow as true peripatetics enduring the heat.[36] Here, as in a problem, you have propositions which must be established one by one, from phenomena revealed by a telescope,[37] if these phenomena are accommodated to those conclusions by means of the axioms of optics, physics, and metaphysics.[38]

But these remarks are made in sport, etc.

. . . . . .

[1] A certain proposition is also the principal matter of the following theses. It was these which, I said in note 37, I wished to demonstrate from phenomena which, in accord with note 38, have been suited to optical, physical, and metaphysical axioms. I shall proceed accordingly below:

I. Phenomenon: On the surface of the moon, when it is regarded as exactly a half moon, the bright part extends, or continues, beyond the terminator and insinuates itself into the other, dark part of the moon.

II. If you apply to this phenomenon unquestioned axioms—that the rays of the sun are rectilinear; that the moon is a spherical body; that the terminator on the moon is nothing other than the boundary of the illumination from the sun, and divides the part of the moon that faces the sun from the farther part, which is turned away from the sun and therefore is not illuminated because of its convexity; and that, given a smooth and perfect globe, the boundary line would be a perfectly straight line at quadrature or a perfectly elliptical one before and after the quarter—it follows absolutely that when the boundary of the illumination is not a perfectly straight line but is broken up with bright teeth jutting into the darkness, the globe of the moon at that point is not perfectly spherical. Those bright teeth are elevations above the dark surface (or, inversely, the dark parts are depressed in comparison to the bright areas near by) with the result that the same rays of the sun which cannot reach the spotted part beyond the boundary (impeded, of course, by its convexity) do reach the bright teeth, as they are raised higher from the center of the lunar globe. The convexity does not remove them from the sunlight (for thus the boundary through them, too, would pass in a straight path) but the altitude of the projections increases as their position progresses into the darkened half of the moon.

III. Phenomenon: The boundary between the bright and dark halves looks rough, like a saw or the transverse fracture of a wooden beam.

IV. Therefore, in that part of the moon which is spotlessly bright, certain points near the terminator rise high. And next to them, also near the terminator, are downward slopes, in alternation. This, in fact, is the definition of roughness. Therefore, the parts of the surface of the moon which shine with pure light are certainly rough.

V. Phenomenon: On the other hand, the terminator passing through the spotted parts of the moon's surface is quite straight.

VI. Therefore, the spots on the moon are part of a smooth and perfectly spherical surface.

VII. Phenomenon: When the terminator passes through the spots, there appear within the illuminated part of the moon certain dark chinks which project outward from the occulted part of the moon. These chinks cut off, as it were, the spots from the spotlessly bright parts.

VIII. Therefore, the rays of the sun illuminate both the bright and the spotted parts on both sides of those dark chinks in the illuminated half; but they do not illuminate the region which the chinks cross.

IX. But, according to II, bright parts are high, spotted parts are depressed. Therefore, the chinks are nothing other than the shadows of the bright points (mountains or cliffs) projected into the spot, as it were, as if onto a level plain or sea.

X. Phenomenon: In the shaded area of the crescent moon, near the terminator, points of brightness can be perceived which, after a lapse of several hours, become brighter until they merge with the illuminated part of the moon on the boundary line. Then it is apparent that these points belong to the bright part of the moon, not to the spots.

XI. Therefore, there must be peaks which rise, from that part of the surface not yet touched by the sun's rays, to such an altitude that they can be touched by the sun's rays. And, again, the whole area around such peaks is higher than the shaded spots on the surface.

XII. Phenomenon: The statements made in I and VII are found to be true as much in the first quarter as in the last, around exactly the same spot. Both boundaries pass through it, each in its own time, but at opposite parts.

XIII. Therefore, the spot, a hollow (according to II) and smooth (according to VI) is surrounded on all sides by bright areas which are high (according to II) and rough (according to IV).

XIV. Phenomenon: In the illuminated part near the terminator appear many dark little sickles, their horns turned toward the terminator. And opposite these dark sickles there are, as it were, reverse sickles, with their extremities touching. These are more saturated with light than the rest of the surrounding area.

XV. Therefore, in the illuminated half, there are rounded hollows, or cavities, that cannot be touched by the sun's rays on the side toward the sun. The remainder of the cavity, which inclines steeply toward the terminator, is more directly exposed to the rays of the sun and is more strongly illuminated than is the rest of the plain outside.

XVI. Phenomenon: One such shining sickle of outstanding magnitude extends its extremities to the very terminator. Opposite it in the illuminated part is a dark gibbous surface, as if cut away from it in a circle. These contrasts of circular light and shade are reversed in opposite quarters.

XVII. Therefore, the shaded half of the moon also contains a huge cavity or pit whose rim, curving toward the sun, casts a shadow into the bottom of the pit; the half of the rim which extends away from the sun into the darkened half receives the sun's rays through a gap or hole in the opposite rim.

XVIII. Among us, the causes which shape the surface of the earth are of two types. One results from intelligence: the cultivation of fields, the erection of walls, the diversion of rivers. The second cause is the motion of the elements. Qualities of the ele-

ments which give rise to motion and transfiguration are humidity
and dryness, hardness and friability. Moisture flows downward toward the places nearer the center of the earth until an equilibrium
is reached. Of the dry materials which lie next to the flowing
waters, the harder ones are more durable and the softer or more
friable ones crumble away little by little. I shall use an obvious
example. Who erected those hills which are scattered over the
fields of Bohemia, where the district narrows in toward the
mountains that form the boundary between it and the Meissen?
If you look out at this row of hills from a high mountain at a
distance, you will say that they are the work of giants—a sort of
burial monument. I shall tell you the real author: It is the River
Elbe which, taking its course down through the mountains, gradually lowered and dug out its channel. It is the passage of time
which, by means of frequent rains pouring down on the rich soil
of the plain, has little by little eroded away the earth and carried
it into the Elbe. Finally, there are rocks which once were underground but now, with the earth removed, are mountains. These
endure by reason of their hardness, while the surrounding soil
crumbles away because of its friability. This is the reason why on
the peaks of most mountains are found masses of rock which will
cause people who have given no thought to the matter to declare
that citadels once were there. This is what has scattered rocks all
over the sandy fields of Silesia. For since the ground is level the
force of rivers is not great; only little valleys or ditches are dug
out here or there by the perennial gushing of springs; the higher
surfaces are spared except insofar as rains erode the fields at the
sides. When the lower areas have been washed away and the
higher areas worn down by the passing of a long period of time,
the rocks which formerly were covered with earth are laid bare.

XIX. Since the intellect is the author of order, and nothing which
is arranged by intellect is unordered or confused unless the intellect
has delegated its authority to instruments different from itself, it
follows that those things which are without order, insofar as they
are without order, arise from the motion of the elements and the
inherent quality of the material.

XX. Since some confusion is perceived on the more visible parts of the surface of the lunar body—some parts being high, some low, some smooth, others rough—there must be on the lunar body something analogous to our elements and the qualities ascribed to them. We may call them by the same names: hard, friable, dry, moist.

XXI. The dark spots on the moon are, therefore, a kind of liquid which, by its coloring and softness, blunts the light of the sun, and which, by its smooth, even flowing around the center of the globe, lowers and levels the surface. It is the mountains which, by means of their own dryness and hardness, also brightly reflect the brightness they receive from the sun, and they rise above the surface of the waters and make the surface of the moon rough by the unevenness of their elevations.

XXII. Phenomenon: There is a difference among the spots, with regard to their darkness; some are blacker than others. For at a distance from the center of the disc, toward the south, is one spot which presents the appearance of an Austrian shield (for there is deep black above and below); but it is divided in the middle by a uniformly wide band which is a little less dark than the area around it but less bright than the bright parts of the moon.

XXIII. In the moon, therefore, the spotted parts—that is, the moist regions—differ in amount of moisture, some being drier, others wetter. Some are analogous to our swamps, some to our seas. For in our swamps, too, grow grasses, reeds, rushes, cane; and everywhere there are also clods of earth, hard and dry and whitish, which more brightly reflect the rays of the sun.

XXIV. Phenomenon: The spots situated around the bisecting line appear, through the best telescope, not dissimilar to the face of a boy disfigured with smallpox if, indeed, the light illuminates this swollen face from either side: for example, from the left, giving the appearance of the first quarter; from the right, the appearance of the second quarter. On such a face all the smallpox tubercles are visible on the side turned toward the light; so also little round spaces, all bright on one side and dark on the opposite side, are seen scattered over the spotted parts of the moon.

XXV. But if the sun's light fell directly on those little spaces, too, it would have to be concluded that there are in fact as many protuberances on the moon as there are small spaces of this kind to receive the light and project a shadow away from the sun. However, we perceive the contrary: that is, those parts of the small spaces which face the sun are shadowed, and the parts that are situated in the region opposite the sun are bright. Hence a contrary shape must be ascribed to these spaces. They do not rise up in little hills, but are depressed in round cavities. Thus it happens that the edge opposite the sun projects its shadow into the bottom of the cavity, but the opposite edge receives the sun's rays more directly and shines the more brightly.

XXVI. Axiom: In the case of those things which are in order, if the cause of the order cannot be deduced from the motion of the elements or from the inherent quality of the material, it is very probably the result of a rational mind. The axiom must be demonstrated by examples. A straight line is an orderly thing; a leaden ball shot from a cannon is carried in a straight line. This motion is not the result of intellect but of the inherent quality of the material. The nitrous material of gunpowder, when set on fire, is forced out and propels the little sphere along the course which it is obstructing; since, therefore, the obstruction continues throughout the length of the iron tube, there is violent expulsion through it in a straight line. The motions of heavy bodies are in such a manner rectilinear that this is a kind of order. A straight line, I say, is in some way or other a property of heavy bodies and especially of light, with rays which are, as it were, a certain body which moves readily. Again, the little house of a snail has the shape of a helix. This is not the result of an architectural intellect but of the inherent quality of the material. For the snail winds itself up into the shape of a cone as winter approaches; over it, thus coiled, flows a viscous fluid which hardens into a shell, and circles are formed in accordance with the number of coils. Again, bees' honey-combs become hexagonal because of the inherent quality of the material of the bodies, when the bees pack themselves in as tightly as possible. On the other hand, the fivefold pattern in flowers has something to do with order, but it cannot be a result of the quality of the material; it must, therefore, arise rather from a formative faculty, which is somehow or other a participant in number and hence in reason. In my book, *The New Star*, chaps. 26 and 27, I discussed whether or not the frequent conformity of many things to one well-defined series could be ascribed to blind chance.

XXVII. Phenomenon: Those cavities in the dark spots of the moon are perfectly round so far as the eye can see; but they are not all of equal circumference. There is one place even where there appears to be a definite order in their arrangement, as of a quincunx.

XXVIII. If we apply the foregoing axiom to these phenomena, we shall arrive at the following conclusions. In general, on the surface of the lunar globe, chance and the inherent quality of the material predominate in causing elevations. Soil is eroded from the rocky subterranean ribs, valleys are washed out, so that mountains rise up, waters flow downward into the low areas marked by spots and there reach equilibrium because of the rectilinear tendency of all parts toward the center of the lunar globe. But in the spotted parts of the moon, the perfectly round shape of the cavities and the arrangement of them—that is, a certain uniformity of the spaces between—is something artificial and the result of some architectural intellect. That hollowing-out in the shape of a circle cannot be the result of any motion of the elements, unless you say that the surface of the moon is overspread with very deep sand, underneath which a hole has been made, and below the crust is an empty place into which the sand flows. Note XXI prevents this conclusion. For moisture occupies these parts, and if an opening were made the moisture would flow down and lay those parts bare, so that what once were spotted regions would become white and shining. Much less possible is it that the arrangement of most of the spots in relation to each other is caused by the motion of the elements.

XXIX. From the foregoing it seems necessary to conclude that there are on the moon living creatures capable of reason that can bring about organization, but not endowed with a massiveness of body comparable to those mountains, since there is no sign of order in the mountains. Men do not make mountains and seas on the surface of the earth either (seldom do we have a Xerxes, seldom a Nero; and not even their works can be compared to the natural ones of mountains and seas), but men do make on earth cities and fortresses, in which order and art are perceptible. Indeed, the surface of the globe seems to have been left to blind chance for this very purpose, in order that in the organizing and adorning of parts of it there might be room for the exercise of reason.

XXX. Phenomenon: If you examine hollows of this sort very closely, and imagine a straight line from the sun through the center of the hollow, six distinct areas appear, three of light and an equal number of shade. It is as if there were cavity within cavity, for the dark part of the big outer cavity has its curved back to the sun, while it turns its bright horns toward the sun and toward the dark part, and its curvature away from the sun. The appearance on the inside of the narrow inner cavity is the same; but light bathes it on the outside toward the sun while its back is curved in the direction of the sun; and on the opposite side, where the horns are turned toward the sun, it is dark. And there is a difference in the outermost rims. In some hollows, this rim is neither brighter nor more visible than the outside region which, I said, belongs to the moon spots, and the curved darkness begins immediately with the same degree of light with which the spots shine. On the side opposite the sun, behind the intense light of the wall which is exposed directly to the sun's rays, a lesser brightness begins and continues through the spotted region. By contrast, in other cavities the outermost rim toward the sun is surrounded by a fine line of very bright light; the edge opposite the sun is surrounded by a thin line of shadow, which marks off the edge from the remaining area.

XXXI. Hence it is demonstrated that from the bottom of the cavity there rises up a hill which has in its center a hollow like a navel, as already in the year 1625 I indicated in the *Shield Bearer of Tycho*, p. 124. Some of the hollows are depressed directly from the surface, others are as if fortified against the outside region by a high wall.

XXXII. The multitude of individual works of art indicates a certain multitude of uses; that is, that there are many users, or that the same user uses different ones at different times. Here, however, reason urges that there is a certain dissimilarity of works analogous to the diversity of times. Thus the ordering in the midst of these numerous things demonstrates the existence of one reason embracing all.

XXXIII. From this axiom and from XXIX we suppose the existence of some race rationally capable of constructing those hollows on the surface of the moon. This race must have many individuals, so that one group puts one hollow to use while another group constructs another hollow. So far as we can perceive the hollows are clearly similar to each other, and are in fact arranged in relation to each other according to a definite plan. Such a plan testifies to a mutual agreement among the makers of the various hollows.

XXXIV. Here add an observation from my *Report on My Observations of Jupiter's Satellites,* bearing the unsuitable title: "Preface to the reader."

Among the observations of September 22, 1622, I find a similar description of what is doubtless the same spot. For again the moon was waning, and the terminator was cutting the small horn off the moon's western edge. My words were: "To the west of the moon's terminator (understand a curve or elliptical line) was perceived something like a shore and high winding projection which cast a shadow into a sea, as it were (toward that horn of the moon already deserted by the light of the sun); for the light followed the shadow in parts of the sea, of the gulf beyond, continuing right on to the boundary line. A little toward the south was something like a luminous isthmus, but with a black point evident in the bright promontory. Beyond the isthmus, a bright mountain was visible in the sea. The horns of that bright projection were extended like promontories. Where the terminator passed through the sea, it was as if cut off by the lower horn; proceeding through another spot, it returned to its own (elliptical or curved) line within the horn." This was in the year 1622, on September 22.

I should like also to write out in full one observation from which I took some of the details mentioned in the foregoing Phenomena, because it still has much to teach us. In it are certain rudiments of the letter whose parts we are engaged in proving:

In the year 1623, on the day beginning at midnight of July 17, I was contemplating the moon from one to two o'clock, using Father Nicolas Zucchi's long-distance glass. Most of the hollows appeared round, but in the upper and lower part of the moon they appeared almost elliptical, following the convexity of the globe as it curved out of sight. There were shadows of valleys in the shape of a horned moon, certain ones obviously caused by the slant of the ellipses, so that from this you could easily discern the roundness of the lunar globe simply by sight. The lower spotted parts were strewn with some bright circles which embraced within themselves the hollows and shadows. These circles were not numerous, however. You might have said that they were swampy or muddy parts of the moon, in which circular ramparts like walls had been erected to shut out the surrounding water. One of these, inclining slightly toward the upper part of the moon, looked very much like a fissure, a little wider in its middle. The distinction between the rims of the hollows and the rest of the lunar body (spotted, or swamp) was not clear, but there was a continuous and consistent faint illumination right up to the opening of the shadow. Most of the larger hollows had in their centers something resembling circular glass panes in windows, that is, individual mounds rose from the bottoms of the individual hollows, but not to the height of the outer walls; and these mounds were depressed again in the center like the navel in a fat belly, or (a more closely related example) the craters of Etna, as their shadow revealed. Still these hollows did not obscure or cling to each other but each stood apart; yet from the lower part of the gibbous dividing line dark little half-moons followed one after another (intersected by one another, as was said) in such a way that in their succession they presented something of the appearance of a dark elliptical arc. Thus was fashioned the image of two dark fissures near to each other. From the division of light and shade they insinuated themselves into the upward bent bright part. And they in turn were interrupted from both sides with bright parts which, as I have said, were crossing over in an uninterrupted course. You might say that there was a very long valley which, winding here and there at the foot of mountains, was, as it were, covered by the mountains when they were viewed from the side. This on July 17, 1623.

These, then, are the observations, these the axioms, with which I shall now demonstrate the individual parts of the letter which I have marked off with numbers.

² That the hollows occupy especially the spots, not the bright parts, pertains to XXIV, and is itself taken from visual experience.

³ That the spotted parts are lower than the bright ones follows according to XIII.

⁴, ⁵ From the spot of complete blackness I deduce seas in XXI; and from the paler black, I deduce swamps in XXII, XXIII.

⁶ If the spaces are measured off there must be agreement among them, as established in XXXIII.

⁷ It is a characteristic of reason to strive toward a definite end. Although it may engage in play to beguile the time, nevertheless the projects provided for play should not be comparable in magnitude to those which have self-preservation as their goal. These works, in fact, are so large that they do not escape our perception from a distance of fifty thousand miles.

⁸ Some of the hollows are surrounded on the outside by a rampart. In the spotted regions, the hollows are depressed, and water enters and causes them to appear black. From this I conclude that the rampart was built to withhold moisture from the outside. Moisture within the rampart, I fancy, had been drained off by those descendants of Endymion, according to methods taught by our Hollanders. In my *Conversation*, p. 17, I fancied the opposite, that the trenches were dug perhaps even for the sake of drawing up moisture from below. But at that time I had not yet observed that the hollows are in the spotted, not in the bright, areas.

⁹ The sun is certainly an enemy to them, and invincible. I wrote about this matter in my *Conversation* as follows:

Since they have a day fifteen times as long as our day and feel intolerable heat and perhaps lack stones for erecting fortifications against the sun (on the other hand, perhaps they do have soil which is sticky like our clay), they have used the following system of construction: They dig out huge fields, carrying out and heaping up the earth in a circle; thus in the depths behind the heaped up mounds, they may bide in shade; and in accord with the motion of the sun they may themselves walk around inside in pursuit of shade. This is for them a sort of subterranean city, their homes being the many caves cut into that circular elevation. In the middle are the fields and the pasture lands, so that, in fleeing the sun, they are nevertheless not forced to depart far from their estates.

I had these thoughts before I noticed that hills rise up from the hollows, and that in the middle is a depression like a navel, and that an outer rampart has been erected around some of the hollows. My line of reasoning was that the inhabitants could more easily acquire extensive shade if they not only dug out a great hollow but also piled up the excavated material on the outside against the sun. What I thought they must have done for this not very certain reason, my eyes and a telescope now prove is done in just this manner; whether against the sun (as note 9 suggests) or against moisture (as in note 8) or against both must be left in doubt.

[10] Once a comparison of lunar and terrestrial peoples is begun, one can make an identical judgment concerning similar things. Since we regard the spotted parts of the moon as cultivated, we shall assume that wild and barbaric bands of robbers inhabit the rough and mountainous surrounding territory. Let these be the enemies of the more civilized against whom the fortifications were erected. But charge to these assumptions the observation in XXXIV, which can be explained in no other way than as a fortification against enemy assault.

[11] The hollows have the form of a circle. A circle is described from a center. The center must, therefore, be visible and furnished with something for measuring the intervals of the circumference.

[12] Equal measurement of such long distances can be achieved only by means of a cord.

[13] Because the diameters of the hollows are unequal.

[14] Since my instrument took in 12' of the 30' of the moon with one view, and since the moon's diameter is four hundred German miles, my instrument took in about a hundred and sixty miles. But the diameter of this particular spot is about a 16th part of the capacity of the instrument; therefore, it extends ten German miles. Hence the radius is five miles.

[15] A circle with a radius of five miles could not be described in a continuous sweep of one leg of a compass unless the compass were held by a surveyor at least twenty miles tall.

[16] It is not enough to bind to the stake a single cord whose loose end is drawn around in a circle. For the cord will sink of its own weight to the ground and will be caught by the hills and rocks and whatever else roughens the surface of the ground. It is necessary for the individual points of the future circle to be distant one from another by no greater an interval than will allow one to be seen from the next. And these points must be designated by means of separate cords fastened to the same stake. Even so the surveyor must proceed from the stake to the circumference with cart loaded in order that he may have a long enough rope for five miles.

[17] From XXIX we gather that the individual members of the lunar race are not comparable in size of body to the lunar mountains, but from XXXIII we gather that these individuals are very numerous. Since the observations suggest enormous works, what could not be done by means of bodily size must be accomplished by numbers. Examples of this sort of thing are the tower of Babylon, the Egyptian pyramids, the very long stone wall in the province of Peru, and the wall that protects the Chinese from the Tartars.

[18] In that great enclosure which has a diameter of ten German miles, a good part of the diameter is taken up by an intermediate gap between the rim and the hill rising up inside. A person who says that the gap is not less than one German mile does not deny visual evidence that it may be greater. Take from this an estimate of the size of their bodies, which, although not comparable to the size of their mountains, is nevertheless much bigger than the size of our bodies, as you can see from their works, which far surpass ours in number. On the basis simply of a comparison of the lunar mountains with ours, I dare to assert plainly in my *Optics,* p. 250:

Plutarch is right in saying that the moon is a body like the earth, uneven and mountainous, and that the mountains of the moon are even higher in proportion to their globe than the mountains of the earth are in proportion to the globe of the earth. But on that we may even joke with Plutarch: on the basis of what happens among us, that men and animals follow the nature of their land or province, there will therefore be on the moon living creatures much bigger of body and much hardier of temperament than we, etc.

[19] Since trenches appear (according to XXV), and these are

artificial (according to XXVIII), it is impossible that they should be made in any other way than by the carrying out of the material. For they do not find any place to bury it in the excavations themselves. It is impossible that nothing can be made out of something, just as art can not fashion something out of nothing.

[20] Grant that this is true of those cavities which are seen not to have a pit fortified on the outside by any bright rim, as was said in XXX, and in the observation from the year 1623 at the end of XXXIV. Moreover, that the material was taken to the inside follows from the appearance of hills rising from the bottom, and from the metaphysical axiom that nothing happens without cause. A height is produced here in the middle of the hollow, therefore it has its cause. And there is no more probable cause than excavation of the material which had formerly filled the nearby ditch visible all around. Nor will it be easy to think up any other cause.

[21] In the case of hollows with bright rims surrounding them, it follows that part of the material has been carried to the outside.

[22] The rampart is double in these because, as I said, the outer rim is the first fortification. But what is carried to the inside becomes a rampart not for the ditch outside (which is not threatened by any peril from within), but for the hollow in the middle of the navel of the hill.

[23] That the moat is very deep is indicated by the abundance of excavated material. Both the exterior wall (which goes all around and is of notable height) and the interior mound (which is likewise very conspicuous) are made from it.

[24] From observation XXVII, and from the method of construction thus far deduced. On the assumption that all points are everywhere equidistant from the stake, and that the moat goes uninterruptedly from one designated point to another, a complete circle will be made by the moat. Other circles therefore will be formed by the exterior and interior walls, as I indicate at number 25. For, assuming the existence of beings to some extent endowed with reason, there is nothing alien to them in this idea of ropes of equal length extended to all points.

[25] [Missing.]

[26] Intellect does nothing to no purpose. But to hollow out the peak of the innermost mound like a navel seems superfluous, for

the wall that rises as a result of that very heaping-up of material from the ditch on the inside leaves a hole in the center, as I point out at number 27.

27 [Missing.]

28 Our earthly mechanics and architecture tell us this, too. It is one of the chief precepts, in fact it pertains even to the chief part of operations, especially in laying foundations.

29 We had swamps on the moon, according to XXXIII. Therefore our agricultural principles are sufficient to lead us to form other conclusions: that certainly if you surround a swampy area with a deep ditch the water both from outside and from within the enclosed area flows into the ditch.

30 If drainage from the area is not sufficient, the material thrown into it from the ditch will make the turf higher and keep it out of reach of the water.

31 In matters guided by reason it is customary for nothing to be done with no purpose. Here is a moat made by rational creatures (according to the hypothesis of my letter) and suitable for receiving water, and it is circular. What good is water going around in a circle if not for sailing on?

32 The length of the very hot day makes it likely that the water will dry up almost every day. In this event, the ditch has another use—for walking around on the bottom of it.

33 The purpose of walking around is to cool off. It is certainly not for recreation or sport (these works are by far too big, a conjecture we accepted among the axioms above) but for the greatest necessity, protection by the shadow of the rampart below from the sun's violence. To make this possible at all times, there is need of a circular moat and a change of position as the sun also changes. This change of position can be made without effort by sailing, or, with more trouble, by foot.

34 Because those who are in the moat have the mound on one side, the outer wall on the other. If, therefore, the sun is in the region of the outer wall, the wall provides the shade.

35 If for them or for those opposite them the sun is in the region of the mound, those who are in the trench betake themselves into the shade behind the mound, or inner wall.

36 For those who are in the crater of the mound, the outer wall

provides no shade. The inner wall provides shade without exertion on their part, because the place is narrow at the bottom around the center, and to cross over from the sun-parched regions to the shaded regions is only a short trip. If the wall does not afford shade for them at midday, when the sun is hanging directly overhead, it is reasonable to suppose that they have caves dug into the steep wall where they can hide and escape the noonday sun.

[37,38] [Missing.]

Thus now I think that everything in the letter has been demonstrated, as I promised for note 37. And this has been done by various and numerous axioms, as I said for note 38. Enough of this letter. With the writing of it I conclude this work of mine, and send the reader to get acquainted with the following work of an ancient writer. [*Note:* The "following work," Kepler's translation into Latin of the original Greek text of Plutarch's *Face on the Moon*, is not included in this translation.]

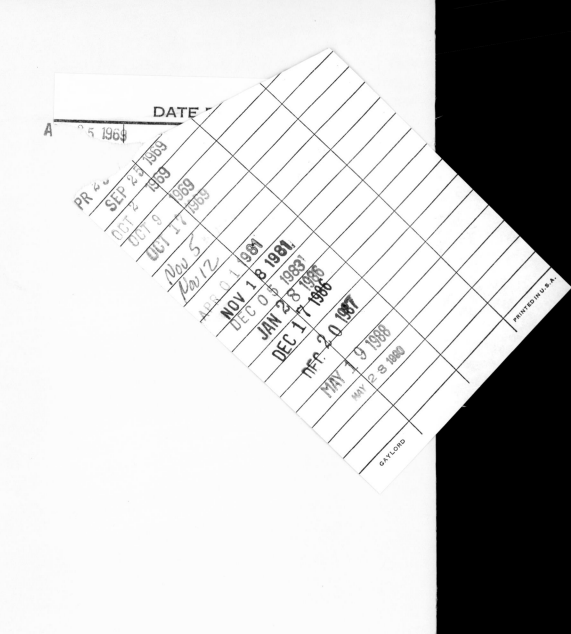

DATE

A   5 1969

PR

SEP 2   1969

OCT 2   1969

OCT 9   1969

OCT 1 7 1969

Nov 5

Nov 12

APR 0 1 1981

NOV 1 8 1981

DEC 0 5 1983

JAN 2 8 1986

DEC 1 7 1986

DEC 2 0 1987

MAY 1 9 1988

MAY 2 3 1980

PRINTED IN U.S.A.

GAYLORD